SEEDS OF SCIENCE

Also available in the Bloomsbury Sigma series:

SEEDS OF SCIENCE

How we got it so wrong on GMOs

Mark Lynas

BLOOMSBURY SIGMA

LONDON · OXFORD · NEW YORK · NEW DELHI · SYDNEY

BLOOMSBURY SIGMA
Bloomsbury Publishing Plc

50 Bedford Square, London, WC1B 3DP, UK

BLOOMSBURY, BLOOMSBURY SIGMA and the Bloomsbury Sigma logo
are trademarks of Bloomsbury Publishing Plc

First published in the United Kingdom in 2018

A catalogue record for this book is available from the British Library

Library of Congress Cataloguing-in-Publication data has been applied for

ISBN: HB: 978-1-4729-4698-0
TPB: 978-1-4729-4699-7
eBook: 978-1-4729-4695-9

2 4 6 8 10 9 7 5 3 1

Bloomsbury Sigma, Book Thirty-four

Typeset by Deanta Global Publishing Services, Chennai, India
Printed and bound in Great Britain by CPI Group (UK) Ltd,
Croydon CR0 4YY

To find out more about our authors and books visit www.bloomsbury.com
and sign up for our newsletters

In memory of David MacKay

Contents

GMO, GM or GE?

A note on definitions. I'm using the terms 'GMO', 'GM' and 'GE' somewhat interchangeably in this book. The first of these is especially problematic. I've used it in the title because it has the highest international recognisability factor, but many scientists I know refuse to use it on principle. What is a 'Genetically Modified Organism' anyway? Your pet dog is genetically modified from the original wolf – otherwise you wouldn't let it anywhere near your kids. All our crops and domesticated animals have been genetically modified from their ancestors to be useful to humans. So are they also GMOs? That's what bugs the scientists: it makes no logical sense to single out anything that has been altered in the lab for special concern and even vilification. Changing genes via laboratory molecular techniques, the main subject of this book, is not much different from conventional selective breeding.

I use the term 'GMO' to indicate the popular debate, and I do not claim it is scientifically valid or even very definable. In fact, the terms 'GM' (Genetically Modified) and 'GE' (Genetically Engineered) are preferable. I've found GM is mostly used in the UK, because in the US it means a large car company. I've used both in the text to avoid repetition.

CHAPTER ONE

UK Direct Action: How we Stopped the GMO Juggernaut

It's three in the morning, and properly dark. But there's still just enough light, cast by nearby street lamps a couple of fields away, to make out the neat rows of maize plants. They are healthy and strong, about shoulder height. Although there is far too little light to see colours, I fancy I can make out the lush dark green of the broad leaves and robust stems. As I ready my machete I'm struck by how, even when it seems pitch dark outside, there is always just enough light – once your eyes get attuned to it – to see by. This theory has only ever failed once: a couple of years earlier, in South Wales, as I traversed a woodland with some other activists protesting against an opencast coal mine. That time it was so dark that I walked straight into a tree. Tonight we're closer to civilisation, somewhere in eastern England; in the deep country certainly, but in that part of England you're always close enough to some human habitation. That's why we keep our torches turned off. You never know who's watching.

I feel a momentary twinge of conscience as I swing back and my machete hacks into the first row of maize. I'm a hobby gardener and have spent time on farms; I don't like to destroy healthy plants. These are admittedly better looking than anything I have ever grown, but then they're genetically modified and therefore to my mind not quite natural. I see this innocent-looking maize as an artificial intrusion, a form of living pollution that doesn't belong in the English countryside. That's why they must be utterly eradicated, I remind myself, as I build up to a rhythm. Slash, whack, chop. Slash, whack, chop. It's surprisingly easy when you get going. The maize plants topple straight over, like trees clear-felled in a forest.

I'm not alone, of course. There are about a dozen of us, spread out evenly across the field, each working a row. You can't be too careful in the near-dark with sharp tools. This would not be a good time to cause an injury. Some of the other activists are close friends; others I hardly know. We have travelled together cooped up in a hired van, a couple of hours on the road, wearing our hoodies and surrounded by metal tools. The dress code is black, or as dark as you can make it. Like all criminals, we don't carry ID, just some spare cash in case of emergency.

It's a funny feeling being on the wrong side of the law. Many people, for good reason or bad, will know what I mean. Suddenly everything is reversed. The friendly policeman is an enemy; you no longer feel a member of everyday society in quite the same way. It is almost as if a veil is drawn between you and ordinary people. You are an outlaw, you carry a secret. You might look normal, but you're not. There are things you cannot say, things that you should not reveal to strangers. On that night in the maize field, as at most times, we are even careful with each other: many people use nicknames or assumed names. Information is typically shared on a 'need to know' basis only. Asking too many questions would lead to suspicion. That fellow activist in the standard uniform of combat clothes and dreadlocks just might turn out to be an undercover cop.[*]

On a different occasion, we were once stopped by police, all piled together in the back of a car with our spades and blades, somewhere in the back roads of Norfolk. The cops made us all get out and stand by the side of the country lane while they wrote down our names and addresses in a notebook. Like a fool I blurted out real details. I spent the next few days in a fog of paranoia, waiting for a knock at the door. It never came, and I still wonder what those police officers must have thought, pulling over this strange car full

---

[*] Several activists I associated with did later turn out to be undercover police, which just goes to show.

of mostly youthful gardeners in the middle of the night. Did they guess what we were up to? Did they really believe our hastily invented cover story about being on our way back from a 'garden party'?

After three-quarters of an hour in the maize field we are making good progress. A good portion of the crop, earlier so vigorous and lush, is now lying wilted and flat. Leaves and stalks, separated from the strong roots that sustained them, are getting trampled into the English mud. But there is still much more to do, and after a short break and some whispered conversations, we redouble our efforts. Slash, whack, chop. Slash, whack, chop. There are car headlights, passing by on the other side of the hedge that runs alongside the genetically modified maize field. Are they moving too slowly, as a patrol car would? We all freeze, but they pass on, humming into the distance. A few minutes later, there are different lights flashing over in the far corner. Have we been busted? We pause again. Then the lights stop – a trick of the night, perhaps. I continue cutting, keeping my eyes on the job, focusing only on the row of maize plants directly in front of me. Slash, whack, chop. Slash, whack, chop.

Then suddenly all hell breaks loose. There are shouts, thuds. People are running everywhere. I hear the unmistakable crackle of police radios. In slow motion, like that recurring nightmare everyone has when you're running in treacle, I try to move, with only one thought: escape. There's an area of woodland on the far side of the field, but between there and where I am now the ground is too open. The best cover is in the tallest part of the maize field itself. Only a few rows away from where I was slashing, I throw myself face down onto the ground. I can taste the earthiness in my mouth as I try not to breathe. I'm not alone – a friend from Oxford is lying next to me, but she keeps trying to whisper something. I bark at her hoarsely to shut up. Then everything is perfectly still, except for the crunching of booted feet through the felled maize as the cops hunt for us. A minute later, they've brought in police dogs. I hear the rapid panting of Alsatians as they charge up and down the rows, sniffing and searching for their silent prey. One comes so close to my prone form that I can almost feel its hot breath as its panting gets closer and closer ... and then further away again as for some reason it passes by.

There's a flurry of barking from a short distance behind us. Someone's been caught. Once they have a hold, police dogs are trained not to let go – there is no point in struggling. Now's our chance. 'Let's go!' I whisper to my Oxford friend, and we break cover, dashing for the safety of the woodland. A 100-yard sprint. There's a barbed-wire fence. Dark clothes are torn, but it doesn't matter. Another wooden fence, then a gate, and we're through, hidden by trees and brambles and steadily getting further and further away from the cops. Once at a safe distance, we hide out in the undergrowth until dawn, then make our way to a nearby railway station. The emergency cash we were supposed to carry is insufficient, but no matter – we vault the barriers and board the first train to London. All the way home I look at the CCTV cameras, in stations and on street corners: are they swivelling to follow me? Have I just delayed the inevitable arrest, followed by the brief appearance in court and then a lengthy jail sentence? I arrive home muddy and exhausted.

This story began, for me at least, in a chilly squat in Brighton in November 1996, three years before the events described above. This particular building was a semi-derelict office block. Bits of concrete and broken glass covered the floors, and the walls were scrawled with graffiti. There was no water or electricity, and even the organisers admitted that only two out of the total 14 floors were usable. It was frigidly cold too: as we bedded down in the rubble in our sleeping bags, it was obvious that no one was going to get much sleep. This was certainly an inauspicious place from which to launch a movement.

The following morning, bleary-eyed and after the customary vegan breakfast (partly salvaged from waste food thrown in skips by nearby supermarkets), we all diverged into different thematic workshops of our choice. Talks were being given on squatting and homelessness, on supporting the Liverpool dock workers who had recently come out on strike, sharing news from the Zapatista revolutionary movement in southern Mexico and updating activists on the situation at the Newbury bypass, where a new road project cutting through countryside and ancient woodlands was being opposed by hardy protesters camping out in treehouses and down tunnels. I headed, along with perhaps half a dozen others, to a side room to listen to Jim Thomas, a professional campaigner then working for Greenpeace, discussing the new subject of 'genetics'. Jim, who I later briefly shared a house with in Oxford, was, like me, in his early twenties, but I remember him as a big, bearded bear of a man, with a warm, almost puppyish enthusiasm that belied his natural air of authority. He was a brilliant activist, with a creative mind, a clever tactical insight and a deep understanding of the issues, for which he was widely respected. Jim circulated black and white Greenpeace leaflets about 'genetic engineering', and his workshop in that draughty building was the first time I heard the ominous word 'Monsanto'. Monsanto. I thought it sounded almost onomatopoeic in its evilness. Almost as if Satan himself had started a company and decided to poison our food.

As Jim Thomas told it, this Monsanto company was genetically engineering crops, starting with soy, in order to patent the new gene-spliced plants and assert an increasing dominance over the global food supply. He was concerned that this technology would intensify the concentration of corporate power and economic globalisation. 'No Patents on Life!' was the new rallying cry. Monsanto was an American chemicals multinational, he told us, and the new engineered foods (the term 'Frankenfoods' only came into widespread use later; contrary to assertions, it was not coined by me) were imminently appearing on European shelves unidentified by any labelling. Perhaps most importantly, these new crops were genetically engineered for one purpose only, Jim Thomas told us: to withstand applications of Monsanto's own herbicide, Roundup. Instead of the more wildlife-friendly farming system we wanted to see, these new unnatural gene-spliced crops would be grown in sterilised fields as part of the worst kind of chemical-dependent monoculture imaginable.

I was hooked. Back at my home in Oxford we already had the perfect vehicle to publicise and drive forward the new anti-GE campaign: a freshly launched activist magazine called *Corporate Watch*. Jim had convinced Greenpeace to give us some old computers they had going spare in their head office, and we camped out either in spare rooms in people's houses or later in a shared activist office space. The magazine was produced and photocopied on the cheap and circulated by post to grassroots direct action groups around the country. I was one of the six co-founders, and the first issue had just been published the month before the Brighton direct action conference, in October 1996. Its black and white cover page featured a cartoon of a boardroom full of corporate baddies discussing their profits while an activist hid under the table with a microphone. Another activist was pictured in a tree outside with binoculars. That was pretty much how we saw ourselves: as investigative gatherers of truth, exposing the misdeeds of the powerful corporations who seemed increasingly to dominate the world. As the first issue editorial page declared: '*Corporate Watch* intends to research and expose the crimes and hypocrisies of those corporations that refuse to act in a responsible manner.' The cover was straplined: 'The earth is not dying, it is being killed. And those who are killing it have names and addresses' – a wonderfully arresting quote attributed to the folk singer Utah Phillips.

Having discovered that Monsanto was the new corporate behemoth killing the Earth with its biotech seeds, I was determined to do what I could to expose and oppose it. Back in Oxford, I wrote a piece headlined 'The Campaign Against Genetic Engineering – It's food, Jim, but not as we know it', which appeared a month later, in December 1996, our second issue. The introduction summarised the whole story: 'In the great global genetic experiment, which is being pursued by chemical and food multinationals in their search for greater profits, we – the consumers – are the guinea pigs. If we let them win their battle to force us to accept genetically engineered produce, reports Mark Lynas, the course of life on planet Earth may be changed for ever.'

It was illustrated by two more cartoons: one showing a befuddled shopper framed by a giant X; the X, conjured up by Greenpeace as a symbolic cross between a chromosome and the X-Files TV series,[1] was to become something of a trademark for the early anti-GM movement; and the second illustration showing a fish-carrot chimera getting overly friendly with an equally grotesque frog-tomato. 'By the time you read this article, you may have already unknowingly consumed food which has been genetically engineered,' I cautioned in the piece. 'Imports of transgenic "Roundup Ready" soybeans from the United States will soon be arriving in quantities that Greenpeace will find it impossible to stop' (Greenpeace had just carried out an action at Liverpool docks on 26 November 1996, trying to stop a cargo ship carrying GE soybeans from docking). 'If we let this happen, the floodgates will open ...' Quoting from newspaper articles and experts about the fearful health and environmental impacts of genetic engineering, I ended the piece with the warning: 'There are dangerous times ahead.'

Corporate Watch was not intended as mere passive information for activists. It was about spurring people to action – direct action. That meant property destruction, theatrics, office occupations, whatever might work in the circumstances to either gain attention or directly alter the situation on the ground. We were briefly notorious in the mainstream press for publishing a booklet containing the names and home addresses of some of the UK's leading company directors, who we felt should be held personally responsible for their presumed crimes. My piece on genetic engineering took the same approach. The third page of the story detailed the main corporate villains, led by Monsanto. 'Monsanto is leading the campaign to push genetically engineered foods onto unwilling consumers' dinner plates,' I wrote, before adding a list of its previous alleged misdeeds, from producing the Agent Orange

defoliant dropped by the US government in the Vietnam War
to manufacturing the artificial sweetener, aspartame. As to
the latter: 'Several studies have linked it with cancer, mood
swings, behavioural changes and seizures.' I did not provide
further details or references to justify these allegations.

An entire facing page was devoted to listing the places and
types of GMOs being trialled in UK fields, at that time
varying from chicory to strawberry to poplar trees. We knew
the exact locations of these GM field trials, right down to
precise six-figure grid references, thanks to the UK
government's 'GMO Public Register'. This was initially set
up to assuage public fear through maximum transparency,
but for us it was a handy tool to locate and then destroy every
GM crop being grown in field trials. One of my friends in
London later maintained a master spreadsheet, with all the
field trials and grid references in separate columns, and simply
ticked them off as each site was 'decontaminated'. It really
was that simple.

As far as I know, my *Corporate Watch* article was – following
Greenpeace's initial impetus – the first in activist literature
aimed at spurring the UK's environmental movement to
action against genetic engineering. The challenge I and my
Corporate Watch colleagues laid out was nothing if not
ambitious. Previously we had tried to stop road-building
projects, or focused on specific instances of ecological harm,
such as opencast coal mining projects or airport runways
cutting into old-growth forests. Now we were trying to stop
the march of an entire technology. We didn't expect to
succeed in stopping GMOs altogether, but we hoped that this
call to mobilisation would trigger an upsurge of direct action
that would at least hold things up for a few years.

We were not disappointed.

The following 21 April 1997 was declared a 'Day of Action'
against genetic engineering. I knew immediately what I
wanted to happen, and which corporation I thought should

be targeted. I had recently discovered that Monsanto had its UK headquarters only 20 miles south of my home in Oxford, in an unremarkable office block nestling in the Chiltern Hills near the town of High Wycombe. I drove over there in my little Ford Fiesta, parked in their company car park and had a look around. The Monsanto HQ was five storeys of red brick, surrounded by open parking areas and with no security gates or fencing to deter easy access to the building. However, there was not much nearby cover; no woodland where activists could hide out before pouncing. We would have to go for the straightforward approach. The only stumbling block would be the keycode on the front doors, but I noted that there was a cafeteria on the ground floor, and that its windows were mostly open.

I got a few hundred quarter-A4 leaflets printed, featuring the same image of the nervous shopper framed by a big X from my *Corporate Watch* article. The flyers were headlined 'GENETIC ENGINEERING – Day of anti-corporate action'. I advertised the time and meeting place, the road round the corner from King's Cross station, a nice convenient central London starting point.* The small print on the flyer said: 'Be prepared for the whole day. Target to be confirmed. This will be a non-violent direct action.' At the bottom it claimed responsibility: 'Produced by the Corporate Action Network' – an outfit that only ever existed in my imagination, but that I vaguely hoped might serve as the direct action arm of *Corporate Watch*. The flyer text under the X image is worth quoting in full. I can't speak for other activists, but it is a good, concise summary at least of my thinking at the time:

> *Genetic engineering involves the transplantation of DNA between entirely unrelated species. It is dangerous and unnecessary – and 60% of your food may already contain genetically manipulated*

---

* I was later embarrassed to be told I'd spelt it wrong: I wrote 'Pancreas Rd', as in the human organ, when it should be 'Pancras Rd'.

products. Huge corporations – like Novartis and Monsanto, aided
by food processors and retailers like Nestlé and Sainsbury's – are
using genetics to engineer a corporate takeover of our entire food
supply. There is still time to stop them.

The next organisational step was to hire buses. That was
easy – I paid for them myself, so the activists I hoped would
turn up on the day would get a free trip. I could afford to hire
buses because I was one of the few people in the direct action
movement with a 'real job', something which I was mildly
embarrassed about and wanted to put to good use. It wasn't
that I worked for an evil corporation: I was the editor of a
small human rights and environment charity website network
called OneWorld, based, conveniently enough, halfway
between Oxford and High Wycombe. But having an actual
job, rather than valiantly subsisting on a combination of state
benefits and scavenging, put me some way down the activist
pecking order. I could not answer immediate action calls and
rush to a roads protest camp facing an eviction, for example,
because during office hours I'd probably be tapping away
quietly in the OneWorld headquarters in a converted garage
in the woods. The activist movement operated an unofficial
hierarchy of commitment, with full-timers living on protest
camps at the top, down to weekenders like myself with one
foot in the scene and the other still anchored down firmly in
mainstream society. On this occasion at least I hoped I could
make a virtue of a necessity and use the money from my job
to help things along.

The big day dawned. I was beyond nervous. Would the
action be a total flop? Would anyone turn up? Would we fail
to get into the building and be forced to stand around like
idiots in the car park, while everyone shouted at me for
insufficient planning? Some good news came early: there had
been a decent turnout in King's Cross, and about 50 activists
were safely on buses and heading out of London on the M40
motorway, destination Monsanto HQ, High Wycombe. With
a couple of other Oxford-based activists I arrived early in my
car and waited at a safe distance. The vague plan was to try

the front door, and failing that the ground-floor windows. Keeping in touch via mobile phone (a rare thing in those days) I asked the buses to wait out of sight until we had the access problem sorted. If those inside the targeted building saw buses rammed full of protesters pull up in full view outside there would of course be immediate lockdown.

In the event the issue was settled in a typically British way. One of the Monsanto employees, unable to resist the instinctive urge to be polite, simply held the door open for us. Thus was the expensive and high-tech keypad security system cleverly evaded. On another mobile phone instruction, the buses pulled around the corner and into the car park. The bus doors hissed open and out leapt dozens of activists, dressed in the most extraordinarily colourful and striking costumes. This was to be the first outing for the Super Heroes Against Genetix (we all enjoyed the acronym: SHAG), masked, caped crusaders wearing the trademark superhero garb of underpants over their trousers. As we Oxforders in our turn held open the doors of Monsanto head office, everyone raced in and bolted straight up the stairs. Within minutes anti-GM banners were being hung from the window of an upper floor, while several other activists had taken over the glass-walled boardroom and had their feet on the table, holding a mock board meeting.

In a matter of minutes the building was ours. Monsanto staffers, following an unseen command from their superiors, filed submissively out of their office, leaving their desks – and more importantly their files – unguarded. I fielded press calls on my mobile while filing cabinets were emptied, their contents either scanned for confidential information or otherwise simply messed up to hinder whatever shady planet-killing work we presumed Monsanto got up to on a normal day. After several hours of negotiation with a few bemused High Wycombe policemen, we all left the building – which was not quite in the same state in which we had found it – and climbed back on the buses, our super heroes with their capes flying behind them and heads held high for a job well done.

It must have been a rude awakening for Monsanto, an unambiguous signal that its attempt to crack the European market with its new GM products was not going to pass unopposed.[2] To my knowledge it was a global first: the earliest targeting of a Monsanto premises anywhere in the world. Many more such actions, some of them much worse from Monsanto's perspective, were to follow.

Following the inspiring lead set by Greenpeace's Jim Thomas, it was my turn to offer workshops. At activist gatherings up and down the land 'Targeting Monsanto' was on the bill, and the news began to spread. At one such meeting, dubbed the 'Big Gene Gathering', two of the UK's most highly respected direct activists, Theo Simon and Shannon Smy, were in the audience. As the two leading lights in the political folk band Seize the Day, Theo and Shannon wrote anthems for the direct action movement. Theo was softly spoken but charismatic and eloquent, with a beguiling trace of a Somerset accent. Shannon, with striking blonde dreadlocks and flowing dresses, played melodic guitar and wrote softer songs with a more emotional touch. I was already a big fan of their music, and later felt privileged to count them both as friends.

But what made Seize the Day unusual among musicians was that they were up for doing much more than just singing about environmental despoliation and global injustice. They were prepared to challenge it directly, putting themselves on the line and risking arrest, prosecution or even injury in the process. I have scarcely ever met braver or more morally committed people. Their sheer audacity was evident from the first time we worked together, when along with three other activists they both stripped naked on the roof of Monsanto's advertising agency Bartle Bogle Hegarty in Soho. Scrawled in red on the banner they held next to their bared bodies were the words: 'Expose the Genetix Cover-up'. I had helped to set up the action a week earlier, by gaining access to the BBH building in my daytime guise as a freelance journalist, sitting through a tedious interview with one of their top executives in order to get a chance to memorise the door key-codes, which I typically failed to do. This logistical hiccup was no deterrent to Theo and Shannon, who on the appointed day found an unguarded stairwell in a neighbouring building and somehow managed to climb out of the fifth-floor windows. As they braved low temperatures and the prying eyes of a large crowd of onlookers, singing and chanting with impressive stamina, I stood down below, warm and fully clothed, and handled the press. This meant writing and faxing around a press release (prepared earlier and dispatched off-site) and giving quotes on their behalf to media on the phone.

According to our press release, the naked protesters were issuing three demands of Monsanto. These were that Monsanto give a 'guarantee that we will suffer no side-effects from eating genetically altered food produced by them, now or at any time in the future', that they 'guarantee that mutant DNA will not leak into our environment', and if they could not, the protesters asked 'will they accept full financial and moral liability for any damage caused by these products to people or the planet?'. The statement accused Monsanto of 'playing God with DNA, and using consumers as guinea pigs'. It continued:

> The naked truth is that the outcome of Monsanto's global experiment cannot be predicted in advance. Already, food which has been doctored with genes from viruses and bacteria is being sold, unlabelled and untested, in our shops. Other new crops being grown in our fields have already begun to spread mutant DNA to related plants in the wild. The natural world is being re-designed for private profit, while agencies like Bartle Bogle Hegarty are paid to keep consumers in the dark about the hazards.*

---

* These quotes come from a OneWorld news story, which I also wrote, thereby quoting from my own press release. Another cardinal rule of good journalism merrily broken!

Talking with the police, I helped to negotiate Theo and Shannon's return to terra firma. They came down triumphantly two hours later, their point made. Monsanto was duly embarrassed, we assumed, and no one was arrested, meaning we were all free to go about organising the next action. Another good result.

One of the strangest experiences I had during this time was as part of the small and secret group that planned what would have been our most daring action of all, had it come off as intended. This was a scheme that none of the participants spoke about for 15 years after the fact, for obvious reasons. I reveal it now in print for the first time.

We had decided to steal science's first cloned farm animal, the world-famous Dolly the sheep.

It's important to understand as background to this story that our objection was not just to the genetic engineering of crop plants and Monsanto. We were against the whole forward march of scientific research in the area of biotechnology and the idea of technological control over intimate life processes such as reproduction. For this reason we were strongly opposed to the emerging technology of animal cloning, and were also queasy about human reproductive advances such as the genetic screening of embryos, seeing it as a slippery slope towards eugenics. We thought that sexual reproduction, as nature intended with all its pitfalls and complications, should be protected from technological intrusion. One proposed action to highlight what we saw as this artificial threat to sexual reproduction was a mass 'shag-in', where dozens, maybe hundreds, of us would have free sex in the open air in order to demonstrate that the natural way was best. I thought this was a terrific idea, and was very willing to be intimately involved in both the planning and the execution. Sadly the shag-in didn't come to pass, as some great ideas never do.

We came pretty close, however, to stealing Dolly the sheep. Dolly had been invented, 'created' we would say, in

July 1996 by scientists at the Roslin Institute, not far from my old university city of Edinburgh. Three activists and I duly took ourselves up to Scotland one autumn day in mid-1998 to carry out our plan. During daylight I posed as an academic researcher and was granted access to the Roslin Institute library, supposedly to carry out research of some sort or other. Once past the front desk, I had free run of the corridors and roamed about trying to find which of the several exterior sheds was the one to contain Dolly. Meanwhile, one of the other activists, who was a surprisingly good voice mimic, pinned up her hair under a colourful floppy hat and adopted a Texas accent. She then pretended to be an American tourist lost on a nearby footpath that just happened to pass close to the same sheds I was trying to gain access to from the inside.

By the evening we had decided we knew the right livestock shed. I had raised some suspicions by being caught rather a long way from the library: 'Yes, sorry, I'm a bit lost. Gosh, that's a big shed over there – what's inside?' Meanwhile our bogus American tourist had done likewise by asking a few too many questions on the outside: 'Is that where they keep Dolly? How NEAT! Which shed exactly did you say?' Nevertheless, long after the sun had gone down, in almost pitch dark, the four of us crept down a country track about a mile from the back of the Roslin Institute. We were shocked, since we were skulking about well after midnight, to walk right into two men coming the other way; fortunately they turned out to be poachers rather than gamekeepers, carrying a brace of pheasants each. There was an exchange of curt nods. 'Evening,' we all said, and went our separate ways.

There followed an hour or more of extremely chilly lying down waiting, under a thorn bush in a field just a few hundred yards from what we hoped was the right shed, while one of us tiptoed out to check that the coast was clear. It was, but the sheds were all locked. Moreover, they were full of sheep. Disaster! As any half-competent shepherd can attest, all sheep look more or less the same. Moreover, cloned sheep, pretty much by definition, look even more the same. After all our elaborate precautions – we never discussed the action plan on the phone, for example, in case of police bugs – the Roslin Institute scientists had outfoxed us by hiding Dolly in plain sight. Frustrated and shivering, we crept back to Edinburgh grumpy and empty-handed just as the first rays of the morning sun crept onto the grassy slopes of the Pentland Hills behind.[3]

Dolly, by the way, only lasted another few years. Overweight and arthritic thanks to being kept inside much of the time precisely in order to foil activists like us, she had to be put down in 2003. Her taxidermied corpse is now on display in the National Museum of Scotland, Edinburgh. Visit her, and you'll get closer than we ever did.

One summer's day in July 1999 I was with several hundred people who had gathered for another 'day of action' in a grassy field near Watlington in Oxfordshire, which just happened to be right next to a 25-acre trial of GM oilseed rape. This was one of the few test sites still remaining in Britain after three years of increasingly focused 'decontamination' efforts on our part. The site was one of a UK-wide scientific programme of government-sponsored 'farm-scale evaluations' on the potential ecological impacts of herbicide-tolerant crops,[4] so we knew that the farmers would be financially compensated whatever happened. We had no interest in waiting to see the results of these experimental fields; we just wanted to get rid of them. Not for nothing was the event called 'Stop the Crop'.

There was a stage set up for speakers, one of whom was Greenpeace's Jim Thomas. The following week, Greenpeace UK executive director Lord Peter Melchett would himself be arrested along with 27 colleagues while levelling another of the farm-scale GM test fields, this time a maize field over in Norfolk. The subsequent legal case had a surprising outcome, when the jury at Norwich Crown Court acquitted the 28 campaigners on the basis that they had a 'lawful excuse' to attack the genetically engineered maize crop because of the threat it posed to the environment. According to press reports, not only was there applause from the public gallery when the Greenpeace activists were acquitted, but some defendants were even congratulated by jury members outside the court afterwards.[5] Clearly the campaign had spread far beyond just Greenpeace and our own freelance direct action efforts. By now it was firmly embedded in mainstream society and popular opinion – hence the verdict by the Norwich jury.

Meanwhile, the movement continued to snowball. Forty local authorities across Britain joined a campaign to remove GM ingredients from school meals, retirement homes and council-run catering services.[6] The supermarket Sainsbury's announced in March it had removed all GM ingredients from its own-brand products, and other retailers quickly rushed to follow its lead. Supermarkets had long been targets of ours: we used to do an action called 'Supermarket Sweep' where we would fill a trolley with food, take it to the tills, and then start loudly asking questions about GMOs. Inevitably a manager would be called, and we would end the action by ostentatiously refusing to buy any products that could not be guaranteed GM-free. I also received a lifetime ban from Marks and Spencer for putting skull-and-crossbones stickers on vegetarian sausages.

A year earlier the campaign had even received royal assent as Prince Charles penned a furious article in the *Telegraph* declaring that genetic modification 'takes mankind into realms that belong to God and God alone', and warning of disastrous effects on human health and the environment. Ex-Beatle Paul McCartney, shocked to discover that his wife Linda's trademark line of meat-free products probably contained GM soy, demanded its immediate removal. Even fast-food chains, including KFC and Burger King, fell over themselves to issue promises to go GM-free.[7] This was no longer just a left-wing political concern. Britain's leading right-wing middle-market tabloid, the *Daily Mail*, launched a

'Genetic Food Watch' campaign, carrying frequent scary headlines inevitably including the word 'Frankenfood'.

Newspapers around the world picked up on the British GM furore. As the *St Louis Post-Dispatch* (a newspaper published out of Monsanto's home town in Missouri) stated in an apt headline, 'Fear is Growing; England is the Epicenter'.[8] As the *Post-Dispatch* reporter put it: 'England has become the center of an anti-genetic engineering food movement that has slowed the march of the biotechnology industry.'[9] It seemed amazing that only three years earlier the entire UK anti-GM campaign consisted of only half a dozen activists in a dingy Brighton squat. Now we had gone global. As the *St Louis Post-Dispatch* suggested, we had created a political earthquake, and the shockwaves were spreading worldwide.

We continued with our direct action tactics at Watlington. The previous fortnight we had occupied a nearby abandoned farm building, turning it into a protest camp and converting its overgrown gardens into 'permaculture' and organic demonstration plots thanks to a tractor-load of imported manure. On the Wednesday before the big event, as media interest about the upcoming protest built up, I found myself in a corner of the GM rape field, facing a camera and with an earpiece in my ear, doing a live TV link-up on nationwide ITN News. On the day itself, there were satellite trucks lining the field: I remember being interviewed inside the Sky News van, with its banks of indecipherable technical equipment, nursing a hangover from a wedding the night before. I was only 26, and this seemed like hitting the media big-time.

Headline-grabbing 'actions' were only the most publicly visible part of the effort. Underlying those were daily hours of tedious work such as maintaining email Listservs, coordinating information-sharing between different groups working in distant cities, and organising gatherings where strategic planning was carried out and new ideas could be floated. I would not claim much personal credit in retrospect for our global success, and certainly far less than that owed to

Jim Thomas and the other dedicated campaigners who together made up the group known as the UK Genetic Engineering Network. My contribution was limited, and focused mainly in the early stages of what later became a broader-based and very effective campaign.

One of our most articulate spokespeople was George Monbiot. George was, and still is, a *Guardian* columnist, a close friend of mine (we lived next door to each other in East Oxford for many years) and a veteran of direct action protests earlier in the 1990s. He had the battle scars to prove it, having suffered a spike through his foot after once being thrown onto metal railings on the ground by security guards on a roads protest site near Bath. Speaking as always without notes, George gave a masterful oration to the Watlington rally:

> We might look to the government to do things about the terrifying spread of these threats to the environment and humankind ... But it has proved again and again that when faced with a choice between what the electorate wants and what big business wants, it will side with big business. We might appeal to international treaties protecting the environment, but they have been overruled by those pursuing coercive trade at any cost. If we do not take responsibility for what is happening no one else will. It is time we stopped wondering what are THEY going to do about it, and started thinking what are WE going to do about it.[10]

If anyone had had any doubts about what was going to happen next, all were removed now. Out came the symbolic white biohazard suits. On went the breathing masks. Up went the rallying cry. Someone let off a smoke bomb, and as music from the stage echoed through the fields, we swarmed across the quiet rural road onto the GM oilseed rape crop – a white army charging into battle against the mutant plants that were stealthily invading our countryside. Almost in a frenzy, along with everyone else I thrashed, pulled up and rolled on the hated plants, even as police helicopters buzzed impotently overhead. A few people were arrested in a fracas somewhere off to the side, but I wasn't among them. I stayed a safe distance away, enjoying the orgy of creative destruction for more than an hour, after which time the majority of the 25-acre field had been levelled. An iconic photograph shows us, all in white hazmat suits, with our biohazard flags flying high, framed by tall oak trees and the picturesque English countryside.* It was an image that defined a movement, a movement then on the verge of stunning success.

Jim Thomas, the Greenpeace campaigner who had started me on this road, later attributed the achievements of the UK anti-GM movement to a number of factors. 'An artful dance of hard work, lucky coincidence, strategy and passion has been behind the gathering momentum of the UK Anti-GenetiX movement,' he wrote. In the town of Totnes in Devon, a kind of UK version of Portland, Oregon, in its multitudes of alternative lifestyle inhabitants (under the road sign announcing 'Totnes' someone affectionately wrote 'Twinned with Narnia'), hundreds – including my own father – turned out to oppose a GM test site that was seen as threatening a nearby organic farm. By 2000 there were multiple groups campaigning against GM crops: Friends of the Earth, Greenpeace and the Soil Association at the professional end, and Genetix Snowball, EarthFirst!, GeneWatch and the Genetic Engineering Network at the grassroots.

By 2002 in the UK there was not much left to destroy. Total field 'decontamination' actions numbered over 70 in 1999, up from 40 the previous year and just a handful in 1997 when the movement first began to gain momentum. On one occasion, 10 GM national seed list trials, essential for the approval of commercial cultivation, were destroyed in the same night.[11] Not only food crops were hit: in July 1999 night-time activists chopped down 50 genetically engineered

---

* You can find it in the David Hoffman Photo Library, photo by Nick Cobbing. I fancy I'm the one with the black backpack, but can't be sure.

reduced-lignin poplar trees being grown at Zeneca Plant Sciences in Berkshire.[12] The trees were never replanted, and the programme was later terminated. In perhaps the most emblematic popular media image of the time, in February 1999 the tabloid *Daily Mirror* printed a doctored photo of a green-coloured Tony Blair with a Frankenstein bolt through his neck. The British prime minister was renamed 'The Prime Monster', under the headline: 'Fury as Blair says: I eat Frankenstein food and it's safe'.

But there was something about this success, even back then, that also made me feel queasy. What kind of victory was it when mass-market tabloids were so gleefully joining a campaign to demonise scientists as latter-day Dr Frankensteins? Should we really be destroying scientific experiments without so much as a second thought? I wasn't sure I agreed with Prince Charles that scientists should be stopped from intruding 'into realms that belong to God and God alone'. It sounded religiously fundamentalist and reminded me of Creationists who tried to stop the teaching of evolution in schools. Was this the sort of thing environmentalists should be involved in? As I moved on to different issues in later years, I began doing my own scientific research, and the doubts grew in my mind. From this seed of science, my misgivings later germinated and spread. Eventually they forced me into a decision that would change my life and bring me into outright and bitter conflict with those who had once been my closest allies and friends.

CHAPTER TWO

Seeds of Science: How I Changed my Mind

My last piece of direct action was a solo mission that has hung around my neck like a millstone ever since. It was one of those things that made me slightly famous in certain circles: enough for random people to say – with surprise, and either admiration or disgust, depending on their perspective – 'So you're the one who ...'

For this reason I still remember the precise date, 5 September 2001. I had returned a few months earlier from Alaska, where I had been carrying out field research about the impacts of climate change, building one of the case studies for my first book on global warming, *High Tide*. In Alaska I had spent time with indigenous Eskimo communities whose lives were being dramatically affected by climate change. In one small Inupiat town on the west coast of the state, Shishmaref, I had seen houses toppling over the sandy cliff. They were no longer protected from the erosion of waves by sea ice, which was forming later in the autumn and melting earlier in spring. Inland, around the city of Fairbanks, I had visited areas where melting permafrost was tilting roads and buildings at crazy angles, and pitching trees into muddy holes in the wilderness called thermokarsts. Whole lakes, I was told, had disappeared, sometimes literally overnight, simply draining into the thawing ground. Ice that had remained frozen since the last ice age was disappearing, and with increasing rapidity. Scientists I had talked to at University of Alaska, Fairbanks and elsewhere had told me that the recent temperature rise was unprecedented, perhaps as far back as 100,000 years. I had also travelled up to the Arctic Ocean coast in Prudhoe Bay, the heart of Alaska's oil industry, and witnessed the incongruity of a state extracting millions of

barrels of oil, all of which when burned added to the rising
global temperatures that were already causing such obvious
problems elsewhere in the same state.

Witnessing these scenes gave me a moral basis for my
protest, which – as I wrote in the subsequent press release –
was carried out 'in solidarity with the native Indian and
Eskimo people in Alaska who are reporting rising
temperatures, shrinking sea ice and worsening effects on
animal and bird life'.[1] I accept that using this information as
the spur for a direct action protest was not standard
journalistic practice, but then I never saw journalism as free
of values, and thought I might as well be explicit about
where I was coming from. And anyway, I was starting to
believe I had science on my side – unlike the target of my
protest, a Danish statistician by the name of Bjørn Lomborg.
Lomborg had recently burst onto the scene with a highly
controversial book called *The Skeptical Environmentalist*,
which contended that most environmental issues were either
flatly wrong or wildly overstated, global warming among
them. He wrote: 'We are not running out of energy or
natural resources ... Fewer and fewer people are starving.
Global warming ... is probably taking place, but the typical
cure of early and radical fossil fuel cutbacks is way worse
than the original affliction ... Moreover, global warming's
total impact will not pose a devastating problem for our
future.'

Lomborg's book was very thick, not least because it
contained over 2,000 references, as well as lots of tables and
graphs. I had been sent a galley proof copy by the publishers,
Cambridge University Press, who for some reason saw me as
a prospective reviewer. Its weight and thickness was helpful
to disguise what I was carrying underneath: a cheap
supermarket sponge cake topped with spray-can cream which
I rather optimistically called – with leaden irony – a Baked
Alaska. In a YouTube video of the action, which took place
in an Oxford bookshop, you can see Lomborg stride
confidently into the shot, remove his leather jacket and
prepare for his talk.[2] Clearly a regular at the gym, he was

well-built, wearing a black T-shirt and with typically Scandinavian blonde hair. Hoping not to attract undue attention, I was wearing my only business suit, which also doubled for use at weddings and funerals. As Lomborg arranged his papers, I sidled quickly up to him with something in my hand. Lomborg was on his feet but unfortunately for him not looking in my direction as he consulted his notes and waited to be introduced. Splat! The 'Baked Alaska' went right in his face, with bits of cream flying everywhere. Utterly shocked and startled, Lomborg reeled back, wiping the cream from his face. Meanwhile, I paced about in front of him and prepared to begin my speech of justification. In the video you can see a few rows of people sitting facing us, in chairs temporarily arranged by the bookshop staff. Everyone gaped but nobody moved a muscle to intervene.

I had expected to be immediately and heroically dragged out of the shop by security guards, which would have photogenically avoided the problem of what to say afterwards – but this didn't happen. As Lomborg – his face and black T-shirt still covered in bits of splatted cream – waited patiently for the teenaged bookshop staff to stop wandering uselessly about and fetch him some paper towels, I had the stage. The spotlight was on me. All I had to do was declaim confidently and eloquently what my protest was all about, how fired up I felt with righteous indignation about Lomborg's dismissive stance on global warming, and how my stance was backed up by rigorous peer-reviewed science.

'That's for everything you say about the environment which is complete bullshit!' I spluttered almost inaudibly. 'That's for lying about climate change! That's what you deserve for being smug about everything to do with the environment …' There followed a pregnant pause. In truth I was already beginning to feel sheepish about what I had just done, especially as Lomborg was still standing there covered in cream. 'Um … sorry but a lot of people have been asking for this,' I said by way of justification. 'Had to happen. A pie in the face for Bjørn Lomborg! Pies for lies! That's what you get for saying lies about the environment.'

Finally – at last! – a member of staff gently asked me to leave, which I proceeded to do with unseemly haste. About two people applauded.

So yes, it was me. I was 'the one who pied Bjørn Lomborg'. And it was an action that probably had far more profound consequences for me than it did for him. I know this because, many years later, I offered him a belated apology. His response was both gentlemanly and rather disarming: he told me to think nothing of it. In particular the episode taught me – as I struggled to marshal a strong case against Lomborg's controversial but exhaustively referenced book – to pay careful attention to supporting evidence. I realised that it wasn't enough to reject his perspective on the basis of my ideological or even moral objections, however passionately they might be held. Whether he was right or wrong on points of fact would need to be determined on the basis of the evidence he presented to buttress his claims, not on how strongly one felt about it. I was given the confidence to throw a pie in his face not so much because environmentalists had been outraged by his claims, but because most of the scientific community had strongly criticised him too. I wanted to 'defend science' by debunking him, so I had buried myself in the two-inch-thick scientific reports of the Intergovernmental Panel on Climate Change. In the process I found a whole new world opening up. I discovered huge numbers of academic journals, which hardly anyone ever seemed to look at, with obscure titles but containing fascinating front-line research on climate change. I felt I had found my new vocation.

In truth it was fairly easy for me to move from protester to writer on climate change because the environmental and scientific communities were closely aligned on the issue. There was enough overlap for me to feel very much at home when I began to attend scientific conferences and worked to familiarise myself with the wider academic literature. When my book *High Tide* was published in 2004, I was worried that someone would take me to task for writing a travelogue, pointing out perhaps that data was not the plural of anecdote or some other related truism. I had already tried to hedge against this by insisting in the prologue: 'Although most of the information in this book is based on the accounts of ordinary people and on my own experiences, its claim to rigour and truth is ultimately founded on the work of hundreds of climatologists, meteorologists, atmospheric physicists and other scientific experts ...' I also made sure that all the 250 references to peer-reviewed science at the back of the book were written in proper academic citation style.

Academic credentials, reasonably enough, don't matter much to environmentalists. A strongly meritocratic movement, its participants rose through the ranks of an informal hierarchy based on creativity, charisma and commitment more than university degrees. Some of the activists I worked with – indeed sometimes the smartest and most effective campaigners of all – had never even finished school, and were formally unemployed or even living on the road. Science, I quickly discovered, was rather different. Sometimes while conducting research I would receive an email addressed to 'Dr Lynas' and would have to admit regretfully that this title could only refer to my father, who had worked as a professional geologist with a doctorate, not me. Not only did I not possess a PhD, but I had no formal scientific qualifications of any kind beyond secondary school. Although as radical activists we were often justifiably sceptical of formal authority structures, I did see that in science, respect for genuine expertise that was based on highly regarded publications in top-ranked journals, pioneering field research or other sources of professional standing made sense. I knew I would never belong to these top echelons of science, but I did at least want its members to respect my work. Either way, all this gave me a strong incentive to make an extra effort to get the science right.

I did have an academic degree, just not a scientific one. I hold an MA in Politics and Modern History from Edinburgh University. The history courses were fascinating but it is safe to say they did not focus much on empiricism. We were taught to believe that the idea of objectivity was more or less a social construction and that an ideological or political perspective was as valid as any notion of universal 'truth'. Now, a decade later as I sat in the dimly lit basement of the Radcliffe Science Library in Oxford, reading scientific conclusions based on actual real-world data felt like a breath of fresh air. It was like drawing back multiple obscuring veils and seeing the world as it truly was for the first time. At university I had learned about the European Enlightenment as a historical phenomenon; now I felt as if I properly understood what it meant for the first time. As I cycled into the library each morning, I had the heady feeling of living through my own personal Enlightenment.

I found I enjoyed being a data nerd at least as much as I had loved being an environmental activist. In the subsequent process of writing my 2007 book *Six Degrees*, I spent over a year slotting conclusions into a giant spreadsheet. These I had sifted from hundreds of different scientific papers published in dozens of journals, varying from geophysics to oceanography to palaeoclimatology, each giving an inkling in their different ways of how the Earth's climate might change with rapidly rising temperatures. Accordingly, *Six Degrees* had over 500 scientific references in the back of the book. My publisher entered it for the Royal Society Science Books Prize in 2008, and to my utter astonishment it won. 'A grim exploration of the implications of global warming has won Britain's most prestigious prize for science writing,' the *Guardian* reported the following day.[3] Although I had used some imaginative and artistic licence in building the narrative for *Six Degrees* – which presented scenarios of climate impacts getting steadily more catastrophic from one through to six degrees as the book unfolded – I was very aware that all my original source material had already been published in journals, and the real research work had been conducted by the scientists drilling sediment cores, running climate models or collecting temperature data, often in difficult or dangerous field conditions. Accordingly I was quoted in the *Guardian* story as saying: 'It's not just an accolade for me, but also for the work of the climate scientists on whose shoulders my writing rests. The book is for a popular audience, and of course it hasn't been peer-reviewed, so to get this accolade from one of the most distinguished scientific bodies in the world means a lot.'

I don't want to appear too naive – romantic, even – about science. In writing my two climate books, I had quickly found that science is complicated. Scientists often disagreed among themselves, sometimes writing rebuttals to papers by other scientists that oozed vitriol. Scientists who had staked out a position with a highly regarded paper often seemed very resistant to changing their minds when challenged by colleagues. Sometimes a headline-grabbing paper that got a lot of cover in top-ranked journals like *Nature* or *Science* would be demolished, or even, very occasionally, retracted by its own author or publisher a year or two later. I knew that, in theory, reproducibility – the ability to repeat a study to check its conclusions – was key, but I also saw that few studies were ever repeated, and that long-running arguments about different conclusions were based on data analysis that required years of training in advanced statistics to begin to understand. Moreover, all this was hidden away in scholarly journals held in the august halls of the Bodleian (to which I was privileged to be given access) or behind paywalls on publishers' websites that charged prohibitive fees of $50 or more for a single download.

When sifting through contradictory material, I learned to trust my intuition as much as anything, and to understand that single papers advancing unlikely conclusions should not be taken at face value on their own. I began to see that scientific knowledge is cumulative: it is built up slowly like a house made of bricks. Sometimes individual bricks need to be relaid or taken out and replaced altogether, but overall the wall generally continues to rise. Only rarely does it get entirely demolished and rebuilt thanks to a paradigm-changing discovery such as plate tectonics or evolution by natural selection. Accordingly the vast majority of those who claimed, Galileo-style, to have overturned a century of scientific research were mostly just cranks.

There were plenty of odd and contradictory papers out there for those who wanted to take contrarian positions. Indeed I and many others had accused Bjørn Lomborg of 'cherry-picking' in our campaigns against his book. In order to avoid this trap myself, I tried to stick closely to the scientific consensus on climate change, and when arguing with climate sceptics in debates or in the media, as a non-expert I felt bound to try to represent the scientific consensus as faithfully as I could. Hence I was a devoted fan of the cumbersome beast that was the Intergovernmental Panel on Climate Change (IPCC), which involved over a thousand scientists publishing monumental reports every seven years assessing the state of the science on climate change. The IPCC was like a lighthouse in the climate storm – albeit one whose beam only came around every seven years.

I received the Royal Society prize for *Six Degrees* on 16 June 2008. The whole way through the ceremony I felt as if I would either wake up from a dream, or someone would tell me that it was a mistake and a different book had won the prize after all (I'm told this is a very common feeling). Yet only three days after the award, when I should have been basking in the glow of approval from the UK's most prestigious scientific institution, I found myself exposed as a hypocrite.

Here's what happened. I was back at home in Oxford when the phone rang. It was an op-ed commissioning editor at the *Guardian*. 'Some government minister has said something positive about GM crops,' he told me (I'm paraphrasing here). 'Can you do us a piece about how he's got it all wrong?' Sure, I replied. I could bash out a quick anti-GMO piece with my eyes shut. I sent it through in less than an hour, and the result was published the same day under the headline: 'GM won't yield a harvest for the world.' In it I asserted:

> *If something goes wrong with a transgenic organism, this raises a whole new category of risk. Traditional pollution – whether of toxins like DDT or radioactive waste – will mix and eventually be dispersed or broken down in the environment. Genetic pollution on the other hand is self-replicating because*

it is contained in living organisms; once released, it can never
be recalled, and possibly never controlled as GM superweeds,
bacteria or viruses run rampant and breed. I am not raising scare
stories here: there are countless cases recorded internationally now
where GM crops have begun to infest supposedly organic or
GM-free fields.

Once the piece had appeared on the *Guardian* website I didn't think much more about it, until in a spare moment I happened to glance at some of the comments underneath. Somewhat to my surprise, most of them were strongly negative. One complained about my 'lack of any kind of scientific knowledge and understanding'. 'This stuff is just Green Party propaganda,' claimed another. The GMO 'fright is Europe's version of creationism', asserted a third. One comment, written by someone using the pseudonym 'Fossil', began as follows:

Lynas, inadvertently or not, engages in the fearmongering about
GM technology that has succeeded in turning the European
anti-GM movement into an object of scorn for scientists the
world over. He refuses to recognize that a gene (and the protein it
codes for) simply is what it is and does what it does. There is no
evil aura surrounding it if it happens to have been found initially
in a virus and then introduced into the genome of a potato. The
only thing that counts is the altered biochemistry of the potato,
which may well be (and in real life usually is) entirely benign.
It is cheap mysticism and rank superstition to imply otherwise.

Although I felt the first prickles of doubt, I knew this must be nonsense. Of course GM plants were contaminating other fields! I decided to do a quick bit of research to prove Fossil wrong, as I knew he or she must surely be, and then maybe add in a reply by myself as author. I felt slightly guilty about not having done this already. The *Guardian* piece had been written off the cuff and published without references, so I went back to the library and began to hunt around among my usual scientific sources. To my increasing consternation, there was nothing much to support my claim that there had been 'countless cases' of GM crops infesting other fields or otherwise spreading damaging 'genetic pollution'. There was plenty about this on the websites of campaigning groups like Greenpeace and Friends of the Earth, but to be on the safest possible ground I wanted to stick to material that was peer-reviewed. There had been occasional controversies, such as whether pollen from GM maize might harm Monarch butterflies or other insects,[4] and a rather bitter dispute over whether GM maize had 'contaminated' native corn in Mexico,[5] but I knew already from my work on climate that basing conclusions on outlier studies that most other scientists took strong exception to risked accusations of 'cherry-picking'.

I decided to take the safe and trusted route of going back to the mainstream scientific consensus, by quoting statements from bodies like the Royal Society or the US National Academy of Sciences to bolster my case. But once again, I couldn't find anything from these sources saying that GM crops were especially harmful. In fact, the august academic institutions I looked at all seemed concerned to say the opposite, that GM crops were most likely to be safe. I found this disconcerting. I remember sitting back in my seat and feeling uncomfortably hot all of a sudden. It was as if a crack had opened up in my worldview, and I didn't know what I would find on the other side. Certainly it was very worrying if real scientists – not to mention the scientific community in general – were on the other side from me on this issue. But if so, this raised some further difficult questions. Was it really possible that not just Greenpeace, but pretty much the entire environmental movement, and indeed polite progressive liberal society in general, had got the GMO issue flat-out wrong? I knew that merely to entertain the possibility was to risk becoming an outcast in the environmental movement, and would certainly affect my friendships as well. On the other hand, if I continued to express opposition to GMOs that was not supported by the scientific community, I could hardly continue to see myself as a defender of science. So that was my choice: I could betray my friends, or I could betray my conscience. Which would it be?

My sense of caution was increased by the unpleasant experience I had already gained of fighting a battle with others in the green movement over the issue of nuclear power. This had begun with a very tentative 2005 piece I wrote as an occasional columnist for the *New Statesman* magazine, the in-house journal of the British left. With the top priority being climate change, I hesitantly wondered whether nuclear power plants, as desperately needed zero-carbon sources of electricity, should not perhaps be kept open longer or even replaced. In response I had instantly been labelled a fraud, a sell-out and an industry shill. Hurt and angry responses had flooded in from friends and readers within hours of the article – which was unhelpfully titled 'Nuclear power: a convert' by the editors – being published online.[6] This was an experience I was not keen to repeat with GMOs. If nukes were a third rail for the 1970s and 1980s generation of greens, GMOs were the same for my own 1990s cohort of environmentalists. This was hardly surprising given how we had fought together in the trenches on the issue. I knew my reputation with my fellow greens was already diminished by the nuclear disagreement, and I was (selfishly, I admit) keen not to do it any further unnecessary damage.

I decided to keep quiet for the moment. However, a year or so later I was sent a book written by the veteran American environmentalist Stewart Brand. Brand had made his name in the late 1960s as the creator of the *Whole Earth Catalog*, which was required reading for any back-to-the-land hippy of the time. His new book was a play on the same title, *Whole Earth Discipline*. It contended that environmentalists such as ourselves had made several key mistakes over the years, among them opposing genetic engineering. Chapter 5, called 'Green Genes', began with this killer line: 'I daresay the environmental movement has done more harm with its opposition to genetic engineering than with any other thing we've been wrong about.' Reading this felt thrillingly subversive, especially as it came from someone who had already spent a lifetime as a green thought-leader. Because his book was so undeniably courageous, I decided to venture a few words of support.

As well as praising Brand's 'splendidly written' book, I tried to be honest about both the nuclear and the GMO paradoxes with my own apology of sorts, writing in the *New Statesman* on 28 January 2010:

> *Although for years I believed in the anti-nuclear cause, I was never an active anti-nuclear campaigner. Genetic engineering, on the other hand, was something I spent years of my life campaigning against. And yet here, too, a science-led assessment of the likely risks and benefits suggests that I was wrong. There is, for example, zero evidence that any genetically modified foods in existence today pose a health risk to anyone ... We cannot criticise global warming sceptics for denying the scientific consensus on climate when we ignore the same consensus on both the safety and the beneficial uses of nuclear power and genetic engineering.*[7]

I'm not sure what I expected, but the reaction was reassuringly muted. I spent the next couple of years back in the library combing the scientific literature as research for my next book, *The God Species*, published a year later. Although GM crops were only a small part of what I ended up writing about, I discovered scientists were mostly talking about the benefits, rather than the harms, of genetic engineering. Apparently GM crops were reducing rather than increasing the use of chemicals, contrary to what I had earlier thought. There also seemed to be ways in which genetic engineering might be able to reduce the use of artificial fertilisers and even help tackle climate change. I wrote a few pages about the issue, but still in rather tentative language.

In the end, it was the issue of scientific consensus that forced me to come off the fence. In 2006 the American Association for the Advancement of Science (AAAS) board had published a strong statement on climate change. 'The scientific evidence is clear: global climate change caused by human activities is occurring now, and it is a growing threat to society.' There was little ambiguity there; it was about as strong as scientific language ever gets. In October 2012, the AAAS board issued a similarly strongly worded statement, this time on the safety of genetically modified foods. 'The science is quite clear: crop improvement by the modern molecular techniques of biotechnology is safe,' it stated, in similar language and with equivalent bluntness to the earlier statement on climate change. The AAAS also referenced the wider scientific consensus among other expert institutions. 'The World Health Organization, the American Medical Association, the U.S. National Academy of Sciences, the British Royal Society, and every other respected organization that has examined the evidence has come to the same conclusion: consuming foods containing ingredients derived from GM crops is no riskier than consuming the same foods containing ingredients from crop plants modified by conventional plant improvement techniques.'

So there it was, clear as day. I couldn't deny the scientific consensus on GMOs while insisting on strict adherence to the one on climate change, and still call myself a science writer. After reading the AAAS position, I felt I had to make a stronger statement, if only to salve my conscience. The opportunity to do so came just a few months later. Although I did not expect it to have much impact, it ended up marking a decisive turning point in my life.

It was 3 January 2013, and I was stepping up to the speaker podium at the Oxford Farming Conference, about to deliver an address to several hundred farmers, politicians and media reporters. The conference was taking place in the Oxford University Exam Schools, a spectacular late-Victorian gothic masterpiece with high carved ceilings and gilded interior. Dressed in suit and tie, I was clutching a printout of my 5,000-word speech. I was more than usually nervous, because I felt instinctively that this was a personal Rubicon moment, something irrevocable that I had been building up to for years and would not be able to go back on. The speech was written out word for word partly because I was worried that I wouldn't have the courage to go through with it otherwise. I certainly wasn't top of the bill: senior government ministers and Prince Charles (via a video address) were also on the speaker roster, and it was them that the media had come for, not me.

If I had thought that the speech would be as widely viewed as it turned out to be, I would also have taken more care over it. I had drafted it hastily, and it contained some intemperate language and needlessly provocative statements. Luckily most of this has been overlooked. For most people it was the very beginning that stood out.

> *My lords, ladies and gentlemen. I want to start with some apologies. For the record, here and upfront, I apologise for having spent several years ripping up GM crops. I am also sorry that I helped to start the anti-GM movement back in the mid-1990s, and that I thereby assisted in demonising an important technological option which can be used to benefit the environment. As an environmentalist, and someone who believes that everyone in this world has a right to a healthy and nutritious diet of their choosing, I could not have chosen a more counter-productive path. I now regret it completely.*

I spent the rest of the lecture attempting to explain how my change of mind had come about, and outlining some reasons why I had got the issue so wrong. At the end I took questions and then went back to my seat. I uploaded the text of the speech to my blog from my laptop, sent out a couple of tweets, and watched the remainder of the conference, relieved that it was all over.

By that evening however I realised that something unusual was happening. My speech had started to go viral. The hit counter on my website was turning like mad. By the following day my server had crashed and the blog had gone offline due to exceeding my allotted bandwidth; the traffic had simply become too overwhelming. The video of the speech posted by the Oxford Farming Conference on Vimeo also began to rack up tens of thousands of views. When my website went back online the speech text was downloaded more than half a million times. Back home, I watched in shock as it bounced around global time zones on social media. I also saw the mainstream media coverage begin to stack up.

Over the next few days the negative responses flooded in too. One widely circulated piece was ominously headlined: 'Uncovering the Real Story Behind Mark Lynas's Conversion from Climate Change Journalist to Cheerleader for Genetically Modified Foods.' Interestingly, the writers of the piece – both from the US-based Organic Consumers Association – seemed to think that my speech had been a significant moment. 'Mainstream media reporters tripped over themselves to get the story out,' they complained. 'Suddenly the Mark Lynas "conversion speech" was major news. "An Environmentalist's Conversion," read the *New Yorker*. "Stark Shift for Onetime Foe of Genetic Engineering in Crops," said the *New York Times* ... The press fawned over Lynas's recantation.'

And now the big question. What the hell was I up to? 'How did a journalist, well-known for his work on climate change, become an impassioned advocate and spokesperson for the biotech industry? And an instant media star in the process? Is Lynas just a slick self-promoter willing to say anything for attention? Or did he sell his soul to the biotech industry?' Which indeed?

Though for some reason they are less easy to recall, I remember that there were lots of emails of support too: I was humbled that people from all over the world offered to translate my speech into their own languages. It was eventually translated by these volunteers, none of whom I have ever met, into Chinese, Italian, German, Spanish, French, Vietnamese, Slovak and Portuguese.

Rather than bashing out instant responses, I tried to keep some sense of perspective. I didn't spend hours sending out tweets or blog posts insisting that I hadn't been bought out by

Big Biotech, as I hoped my record spoke for itself – and anyway, it was obvious that no denials would convince my detractors. This meant that I didn't correct the record either when pro-GMO pieces over-egged my role in the early anti-GMO movement for their own purposes. According to the *Australian*,[8] I was 'one of the first leaders of the anti-GM movement in the mid-90s', while the *New York Times* had me as someone who 'once helped drive Europe's movement against genetically engineered crops'.[9] In one television interview I was described as 'one of the godfathers of this movement'. This label stuck, and even got more extreme over time. By 2015 one article was headlined: 'Why the founder of the anti-GMO movement converted to the side of science.'

When many of those I had worked with in the UK during the 1990s, including Jim Thomas, Theo Simon and various other senior people from Friends of the Earth and Greenpeace, signed a statement saying that they did 'not recognise Lynas's contribution as being significant in the ways it is being represented',[10] I found if anything that I rather agreed with them. But there didn't seem to be much I could do about it.

One of my most memorable media encounters took place less than a month after my Oxford speech, on a famously confrontational BBC World show called *HARDTalk*, hosted by Stephen Sackur.[11] Sackur was disarmingly pleasant as we miked up and chatted in the green room. But as soon as the studio cameras rolled he adopted a very different persona. As he might with errant politicians, Sackur focused on my inconsistency, my fallibility for having got an issue wrong and the shame of later being forced to admit it. 'Let's start with your very high-profile recent renunciation of your pretty much lifelong commitment to campaign against genetically modified food production,' he began, fixing me with his trademark withering glare. 'Is it fair to summarise by saying that you have concluded that everything you used to think was entirely wrong?'

The rest of the interview proceeded in pretty much the same vein. As I tried to come up with reasonable responses, Sackur pounced on each admission. 'So that, it's fair to say, leaves your personal credibility in shreds,' he suggested at one point with mock innocence, fiddling with his pen in intimidating prosecutorial style. This pushed me into admitting that I was 'not proud' of having been part of a campaign that had 'done real damage'. 'You're ashamed of the entire approach you took, your complete lack of intellectual rigour,' he persisted. Once again, I said, I had already apologised for destroying GM crops and – choosing my words more carefully this time – for 'having made a contribution to starting this movement'. Sackur went in for the kill. 'And if you were so wrong, so incompetent, so shallow in the past, why should we believe you are any different now?' I could only think of one response to that. 'All the more reason for me to change my mind,' I said with a mental shrug.

What Sackur didn't seem to realise was that unlike his usual targets – politicians, business leaders and the like – I had no problem with admitting my U-turn. In fact, that was the whole point. As the economist J. M. Keynes is supposed to have said:* 'When the facts change, I change my mind. What do you do sir?' While changing your mind may be a sin in politics, in science it is supposed to be part of the job description. So I felt that I could make a case that not only was changing my mind a reasonable thing to do, it was the right thing to do on the basis of the actual evidence.

In case this all seems a bit starry-eyed about science, let me offer an example from personal experience where this did in fact happen, in a high-stakes experiment about this very

---

* This is probably an attributed quote.

issue of GM crops, and at substantial risk to the reputations of the scientists and the institution involved. This regarded an important piece of experimental work being carried out by scientists at Rothamsted Research, a publicly funded plant science centre based at Harpenden in the south of England. In early 2012 the institute established an open-air field trial of genetically modified wheat (the first in the UK for many years), sown behind a high fence and protected by 24-hour security. The aim of the research was to test the hypothesis that wheat with an added gene to express an aphid alarm pheromone would repel aphids, thereby reducing the need for chemical pesticides. The stakes were high because a group of anti-GMO protesters had vowed to destroy the test site before the experiment could yield any results. In response the scientists released a passionate YouTube video appeal and took to the media to plead their case that their potentially chemical-reducing wheat experiment be allowed to proceed. I had a small role in helping the Rothamsted team behind the scenes, encouraging them to have the confidence to leave the lab and talk to the public honestly about their work.

One of the scientists, Toby Bruce, made the basic case in the video.[12] Addressing the camera directly, he said:

> *We have developed this new variety of wheat which doesn't require treatment with an insecticide, and it uses a natural aphid repellent which already widely occurs in nature and is produced by more than 400 different plant species. We have engineered this into the wheat genome so that the wheat can do the same thing and defend itself. Are you really against this? Because it could have a lot of environmental benefit. Or is it simply you distrust it because it's a GMO?*

Another Rothamsted scientist in the video was Janet Martin, who asked quite reasonably: 'You seem to think, even before we've had a chance to test the trial, that our GM wheat variety is bad. But how can you know this?' She paused and uttered a weary, unscripted sigh, before continuing. 'It's clearly not through scientific investigation, because we've not even had a chance to do any tests yet. You state on your website that there is serious doubt that the aphid alarm pheromone being produced by this GM wheat is going to have any effect anyway, and you could very well be right. But if you trash the trial none of us are ever going to know, are we?' The video and associated public appeal seemed to strike a chord. Thousands signed a petition set up by the pro-science campaign group Sense About Science, and social media buzzed with the hashtag #DontDestroyResearch. Press coverage was largely sympathetic to the predicament of the scientists, a striking change from a decade earlier when the activists seemed to be leading the agenda. When the day came for the promised action, on 27 May 2012, too few protesters turned up to break through the fence and the wheat experiment was able to continue right through to harvest.

Having invested so much staff time, millions of pounds and a substantial slice of its institutional reputation in fighting a campaign on behalf of the experiment, it might therefore have been seriously embarrassing for Rothamsted when the results came in – they showed, pretty conclusively, that the trial had failed. There was no denying it: the GM wheat did not repel aphids as expected. But Rothamsted, to its credit, didn't try to bury the bad news. The team published an open-access paper in the *Nature* journal *Scientific Reports* stating straightforwardly that 'field trials … showed no reduction in aphids',[13] and just for good measure accompanied it with a press release headlined: 'Scientists disappointed at results from GM wheat field trial.' This press release stated bluntly: 'The data show that the GM wheat did not repel aphid pests in the field as was hypothesised and was initially seen in laboratory experiments conducted by scientists at the Institute.' Toby Bruce, one of the scientists in the video, was also quoted. 'In science we never expect to get confirmation of every hypothesis,' he said. 'Often it is the negative results and unexpected surprises that end up making big advances – penicillin was discovered by accident, for example. If we knew the answers to every question before we started, there would be no need for science and there would be no innovation.'

I think the nicest irony of the whole saga is that Rothamsted's wheat trial ended up proving the activists right and the scientists wrong. But it did so not through ideological assertion but via rigorous empirical evidence. A finer example of the primacy of the scientific method can scarcely be imagined.

CHAPTER THREE

The Inventors of Genetic Engineering

A few days after my Oxford speech I received an out-of-the-blue email from Professor Nina Fedoroff, a scientist who had herself done pioneering work in the field of molecular genetics. She was also then president of the American Association for the Advancement of Science, and had helped to put together the AAAS statement about the scientific consensus on GMOs that had encouraged me to speak out publicly. Fedoroff was also former science advisor to Hillary Clinton during the latter's time as US Secretary of State. Not surprisingly I was somewhat in awe and read her email eagerly. 'Welcome to the science side,' Fedoroff began archly. 'You may be in an uncomfortable spot, but please hang in there. It's a whole lot less uncomfortable than the spot Galileo found himself in.'

Fedoroff was concerned that since I was suddenly in the spotlight, I should be as well informed as possible. 'You're not saying anything we haven't been saying for years,' she continued bluntly, 'but you're getting a lot of attention – use it.' She then generously offered to be my informal science advisor. I sent back a few queries; Fedoroff suggested I read *Cell and Molecular Biology for Dummies* (I did, and found it surprisingly useful). I also read Fedoroff's own 2004 book, *Mendel in the Kitchen: A scientist's view of genetically modified food*, which related some of the fascinating wider history of genetics. The Mendel referenced in the book's title was Gregor Mendel, the nineteenth-century Czech monk who discovered the mathematical rules of genetic inheritance through experiments with cultivated pea plants. The story continued to the 1953 breakthrough of Francis Crick and James Watson, the two young Cambridge scientists who together with Rosalind Franklin and Maurice Wilkins figured out the chemical structure of the DNA molecule, the now famous double helix.

Fedoroff's history continued into the more recent era, to the scientists who discovered how to combine DNA from different sources into an existing genome. Her book rekindled my instincts as a historian, and I decided to look more closely at the actual invention of genetic engineering as a technology. What I found was little known but deserved to take its place as a classic tale of scientific discovery. Three teams across two continents, beginning in the early 1970s, raced each other for over a decade, starting with blue-skies pure science and ending in a furious battle to be the first to bring the new applied technology of genetic engineering to world agriculture.

It was late February 2016 and the 'father of genetic engineering' was waving to me through the crowds at the Eurostar terminal of Brussels Midi station. Professor Marc Van Montagu, in his old-fashioned and understated Belgian way, would never be so crass as to use that term himself. But later over a seafood dinner with his former dentist wife Nora, she could not help herself – she clearly felt her husband was not getting the recognition he deserved. 'Who is the father of genetic engineering?' she exploded, in response to my gentle prompting. 'It's him of course!' She pointed across the table as Van Montagu sat with lips pursed, holding his napkin and embarrassedly contemplating a half-eaten fish. Nora insisted that he should write a book about the experience, but Van Montagu had his excuses ready: 'All the documents were lost' … 'It was too long ago.' As I watched I got the sense this was a well-rehearsed argument of the sort familiar with couples who have been married a long time, in their case 59 years. Van Montagu wouldn't be drawn, but his wife's point was clear enough: he was then 82, and he needed to tell the story or risk it being lost for ever.

Marc Van Montagu and Nora Podgaetzki met at university when they were both only 20. They had both experienced the trauma of the Second World War as children, though Nora's story is the more extraordinary. As a Jewish child, she was 'hidden' in the Belgian town of Ghent for two years during the Nazi occupation between 1942 and 1944. Amazingly, she continued to go to school, and the whole class shared the secret that they still had a Jewish girl attending lessons. No one in the playground, not even the youngest child, broke the silence and told the Germans. Everyone knew by then that being Jewish meant deportation east to a concentration camp and probably death. Nora was hidden by a Belgian family, by parents who had children of their own. The father was a professor, she told me. He was also running an incredible risk. If the fact that he was sheltering a Jewish girl had become known, the Nazis would probably have sent not only Nora, but also the professor and his children to the death camps. Nora's parents also remained in Ghent, and had to stay in hideouts for the entire duration of the Nazi occupation. On one occasion Nora and some other children were taken to play in the garden of a house on the other side of town. She didn't know it at the time, but her parents were peeping out from behind the curtains. That was the only time they set eyes on her for two years.

Van Montagu's childhood story is also a testament to how different Europe was in the early twentieth century, and how dramatically things have changed. He was born in 1933 in a working-class neighbourhood of Ghent. Van Montagu's mother died during the birth. As he wrote later, 'Death of either the mother or the newborn was very common in those days. My mother was the only survivor of my grandmother's nine pregnancies.' Left motherless and without siblings, Van Montagu was raised by his maternal grandmother and her sisters, 'surrounded by lots of love and attention, being the only child of my generation in the whole family'. Ghent at the time was a textile town, 'large factories surrounded by a network of dead-end alleys with small working-class houses'. Pre-war Belgium was poor. 'Most houses did not have running water; there was a central tap in the street. Some even had common toilets in the middle of the street. Light came from petroleum or gas lamps; few houses had electricity.'

The young Van Montagu was quickly convinced that he must work hard at school if he was to avoid the drudgery of factory life. 'The factories were dark and noisy, and clouds of cotton dust would be floating around the spinning machines. They were so frightening and convincingly repulsive that I felt I never wanted to be obliged to work there.' Even this picture of industrial hell represented an improvement in working life from his grandmother's time: working hours had been reduced from 12 to 8 hours a day over previous decades. Moreover, in his grandmother's youth 'fifty per cent of the workers were children, many younger than ten years of age. They were important for sliding under the machines and knotting the broken threads.' Clearly no one worried much then about health and safety.

Van Montagu grew up very aware that improvements in factory labour conditions had not come about by accident. They had been fought for by generations of working people active in a powerful socialist movement, of which Van Montagu's great-grandfather-in-law was one of the founders in 1870. As we sat talking in his spacious and warm living room, surrounded by eclectic art from around the world and sipping coffee, Van Montagu proudly scrolled through web pictures of his illustrious working-class ancestor Edmond Van Beveren, who is memorialised in an impressive grey-steel statue at Ghent University.[1] Van Beveren is remembered today as the father of the Belgian Labour Party, and the young Van Montagu grew up in a fervently socialist household. 'May 1st (our Labour Day) was the most important holiday of the year. Never would we miss the parade, and the whole day we would sing militant songs.' Later at university Van Montagu would rise to become a national leader of the youth wing of the party, and would remain a card-carrying socialist his whole life.

As we talked through the fading afternoon, I was struck by a profound historical irony. I and so many anti-GMO campaigners, motivated by a leftist concern about genetic engineering being pioneered by big American multinational corporations, laboured in ignorance of the fact that one of the most significant contributions was made in the heart of Europe by a lifelong socialist working in a public university.

After the war and while at high school, the young Van Montagu, increasingly fascinated by chemistry, ran a gas line into the attic of his house and set up a laboratory. With only a single coal stove in the house, 'conditions were a bit harsh' in winter, but 'the heat of the Bunsen burner was sufficient', he recalled stoically. The rest of his time, while not conducting experiments, was occupied with reading books. By the end of high school, Van Montagu remembered, 'I was already resolute to study the chemistry of living organisms, biochemistry as it seemed to be called.'

This is important for what was to come later. Throughout his career Van Montagu was to see his work with living cells through the eyes of a chemist, rather than those of a biologist. As he reminded me, the cell is the lowest constituent of life that is actually alive. Everything that happens within the cell is simply chemistry. DNA, RNA (ribonucleic acid), amino acids, lipids, proteins; all are inanimate molecules, yet somehow they become arranged in a dynamic, ever-changing way that overall makes up a living cell. Post-Darwin, biologists tended to think about organisms or species as the main units of interest. Van Montagu saw things differently. Rearrange the order of inert molecular constituents within the cell and you could rearrange the organism itself, he realised. Accordingly, he ignored advice from a trusted professor to go into pharmacy: 'I was afraid I would end up in a chemist's shop.' Spending his life selling aspirins over a chemist's counter in small-town Belgium was certainly not Van Montagu's ambition.

At the University of Ghent, Van Montagu's interest in cell biology grew. It was 1952, just months before Watson and Crick were to publish their seminal paper on the structure of

DNA. With so little known about how DNA and cells worked, even lowly undergraduates like Van Montagu felt as if they were pushing at the frontiers of science, he told me. Only about a decade had elapsed since many experts had insisted that endlessly complex proteins, not simple, repetitive DNA, must be the molecular unit of heredity in living organisms. We now know that it is the sheer repetitiveness of DNA, with its four 'bases' – A, T, G and C – that enables DNA to store information. Different sequences code for different amino acids, which then form together into proteins, which in turn do most of the work in building and maintaining the living cell.

By the end of the 1960s, Van Montagu was working on what happens when DNA goes wrong. In animals, we call malign DNA mutations that lead to proliferating cells 'cancer'. Plants get cancerous tumours too, which manifest in big knotty growths called 'crown galls'. These galls are not generally caused by spontaneous mutations in DNA, however. Instead they tend to be associated with a bacterium called *Agrobacterium tumefaciens*, so named because of its ability to induce tumours. Other scientists had already proposed that something – a 'tumour-inducing principle' – must be passed from the bacterium to the plant in order to generate these crown galls. Van Montagu's contribution, in a series of pioneering experiments and subsequent scientific publications, was to show that this 'tumour-inducing principle' was DNA, and that *Agrobacterium* was therefore a natural plant genetic engineer – splicing its own DNA into the cell of the host plant.

Few scientists work alone, and Van Montagu's long-term collaborator was Jozef (Jeff) Schell, whose contribution was every bit as significant as Van Montagu's own. Schell died in 2003 at the relatively young age of 67. The journal *Plant Physiology* in its July 2003 issue made an exception to its usual rule of not publishing obituaries in recognition of 'Schell's enormous contribution to plant science'. In 1998 both Van Montagu and Schell were awarded the Japan Prize from the Science and Technology Foundation of Japan; in his acceptance lecture Schell gave one of the most cogent explanations of plant genetic engineering I have read anywhere:

> *Two scientific breakthroughs underlie genetic engineering in plants. First was the development of recombinant DNA technology which made it possible to isolate individual genes from any organism. The second was the discovery that there are bacteria in the soil, Agrobacterium tumefaciens, which transfer genes into plants. This was the first documented instance of plant genetic engineering in nature.*

When I asked Van Montagu in his Brussels living room who first had the idea of harnessing this newly discovered bacterial process as a way of splicing useful genes into plants, he batted the question away: 'I think it is obvious!' Of course brilliant ideas often seem obvious after the fact – the first mention I could find in writing was in a typed 1975 manuscript by Jeff Schell, which tantalisingly left this 'obvious' idea to the very final paragraph. After explaining the technical specifics of how *Agrobacterium tumefaciens* causes crown galls in plants, Schell wrote: 'The possibilities for genetic manipulations with plant materials should be obvious … Various genes of interest could be introduced in this plasmid [the piece of circular DNA transferred from the bacterium to the host plant's genome] and thus possibly be stably transferred to the plant cells.'[2] Exactly how to do that was far from 'obvious' however, and the task of doing so would consume Van Montagu and his collaborators for the next eight years.

As Schell had noted, the pioneering work on recombinant DNA that helped make plant genetic engineering possible was being conducted around the same time. In 1971 Stanford University's Paul Berg had created the first recombinant DNA molecule in an experiment that led to him being awarded the Nobel Prize in chemistry in 1980. This wasn't strictly speaking a living organism: Berg had spliced together DNA from a monkey virus and a plant virus, and viruses can't really be considered alive. But the point was it worked. With the right mix of enzymes, DNA could be forced to recombine from different, unrelated sections, making a genetic chimera. Within a year Berg's Stanford colleague Stanley Cohen, in collaboration with Herbert Boyer at the University of California San Francisco, was inserting recombinant DNA plasmids (circular bits of bacterial DNA carrying two different antibiotic resistance genes) into *E. coli* bacteria and watching them be replicated just like the rest of the bacteria's genome.

Cohen and Boyer were also the first to break the mythical species barrier, by shuttling DNA from a giant clawed toad (a 'model organism' that biologists often tended to have sitting around in the lab) into *E. coli*. Clearly if toad genes could be isolated and stably transferred to *E. coli*, so could human genes. By 1978 Boyer was working on moving human insulin genes into *E. coli*, and by 1982 human insulin produced by bacteria (and later yeast) via recombinant DNA was on the market. This relieved the previous shortage of pig-sourced insulin and provided a lifeline to diabetics worldwide. Commercial biotechnology was born.

Plants are different from bacteria, however, and moving genes into them presented a greater challenge. Bacteria have relatively permeable cell membranes, so getting recombinant DNA sections into them was easy. But plant cells have a solid wall, which DNA cannot cross, and the additional barrier of a nucleus on the inside. Hence the interest in *Agrobacterium*, which had naturally evolved a way to get its plasmid genes into plant DNA via tiny tubes. Later, genetic engineers would also try using ballistics – literally a gene gun – to fire DNA-coated particles into the plant cell nucleus.

There were plenty of sceptics about Schell and Van Montagu's achievements at the time. One of these first came to Van Montagu's notice when he received a transatlantic phone call, a rare thing in those days, out of the blue. It came from one of the leaders of a University of Washington team of scientists in faraway Seattle. The caller was an up-and-coming biochemist called Mary-Dell Chilton. She was characteristically blunt. Van Montagu's results, she said, were 'nonsense, total nonsense'. There was no way that *Agrobacterium*'s genes were actually present in the plant cell. And she, Mary-Dell Chilton, was going to prove the cocky Belgians wrong. With that, she hung up.

'We found ourselves in the role of iconoclasts,' Chilton wrote later about her competition with Van Montagu's Belgian group. 'It was essential to know what could be believed.' Chilton's refusal to take something on trust, even from a fellow expert, seems impolite but is of course in the proper tradition of science. It reminds me of the Royal Society's motto, *Nullius in verba* (which roughly translates as 'take nobody's word for it'). Accordingly, Chilton was determined to demolish Van Montagu's theories about *Agrobacterium*'s role as a genetic engineer of plant cells.

When I pressed Van Montagu for an opinion on this, he showed no trace of resentment. 'That's the way science has to be done!' he said, beaming. 'I think it's a very good example, and it's all in her honour that if she says she doesn't believe it, then you do good science and see who is correct. When you don't believe, you are angry about something, you analyse.'

Chilton's approach to Van Montagu's work was entirely in character. She was raised largely by her grandparents in North Carolina, and her school scores were always so high that teachers at first thought she must have cheated. In retrospect, she was always cut out to be a research scientist: 'I liked chemistry because this was a topic in which I could ask a question and they didn't know the answer yet. I could see the research edge in chemistry,' she remembered. Molecular biology as a field was only 'just invented at the time … I wanted to know how all of this worked. And it was easy to ask questions to which we really didn't know the answers.' It didn't take her long to zoom in on DNA in particular, which she noticed seemed to have the property of self-correction. It could fix a mistake, reverse a mutation and put itself back in order in an apparently spontaneous way. 'And that looked as though it were a molecule with brains. I wanted to know how those brains worked.'

Chilton admits that 'I've been a very wilful and competitive person all my life', and as a young researcher she was not about to sacrifice her work for the sake of becoming a mother. This didn't mean she would remain childless: she would simply do both things at once. 'I actually talked to my little boys about that. I told them that if they wanted to have a mother who was happy, that she was going to be a mother that worked on her science and that was the way it was going to be.' Chilton got the *Agrobacterium* bug when one of her own students presented a paper on the little-known plant cancer-causing bacteria. 'I was fascinated,' she recalls. 'Apparently people thought that *Agrobacterium* might be moving genes from itself into plant cells.' This sounded intriguing, but Chilton's sceptical instincts were triggered as soon as she checked the story out. 'As I looked at the publications, they were terrible. They hadn't done the proper controls, so you couldn't really believe the results. They hadn't done it right.'[3] This was not necessarily a comment on Van Montagu and Schell's research. Others were working on *Agrobacterium* using what might now be considered flaky experimental protocols.

Chilton admitted, when I interviewed her on the phone from North Carolina, that her later interactions with Van Montagu's team were 'fiercely competitive'. The two teams 'didn't work together at all', she told me. But competition can be a driver of scientific progress just as cooperation can be. As Van Montagu related, Chilton was driven by her determination to disprove the upstart Belgians. She thought that the theory that *Agrobacterium* was splicing its DNA into plant cells was (in Van Montagu's words) 'just hype, that the facts were wrong'. Testing that proposition was not a simple undertaking, however. So, in 1977, the University of Washington team in Seattle embarked on a major 'brute force experiment involving everyone in our lab', as Chilton later described it. Weekend plans were cancelled and family lives thrown into disarray. 'We all did whatever had to be done next. I have never experienced such completely committed teamwork in my entire career, before or since.'

As is often the case with major breakthroughs, even decades after the event Chilton can remember the exact moment of truth. In a sprawl of printer tape, 'I suddenly saw that the T-DNA [transfer-DNA] was in the tumour cells.' Chilton's team had found the precise section of genetic material transferred from the bacterium to the host plant. By way of confirmation, the subsequent scientific paper included a photograph of the matching banding patterns made by the identical DNA sequences extracted from both the plant cells and the *Agrobacterium* plasmid. Chilton and her co-authors wrote: 'Our results suggest that the tumour-inducing principle ... is indeed DNA, as many investigators have long suspected.' Therefore, she continued, 'Crown gall tumours can, in light of our findings, be viewed as a feat of genetic engineering on the part of *A. tumefaciens*.' She added, echoing Schell two years earlier: 'If the foreign DNA in the tumour cell does function this way, the potential applications ... in future genetic engineering studies in higher plants are apparent.'

So Mary-Dell Chilton had not debunked Van Montagu and Schell. Instead she had proven experimentally that they were right. The Ghent-based team was forced to concede defeat in the race to prove the facts. As Van Montagu wrote later: 'Unfortunately, we lost the battle to publish first, and the only record of our result is a talk given at a symposium in 1978.'[4] The usurpers across the Atlantic in Seattle had pulled ahead in the genetic engineering race.

In 1979 Chilton made a personal decision that would have profound consequences for the new technology she was helping to pioneer. She moved with her family from Seattle to Washington University in St Louis, to begin work in a research laboratory based just five miles from a company that was already showing a great deal of interest in her research: Monsanto. One Monsanto executive, Ernest Jaworski, had for years been following the work of both the Chilton and Van Montagu teams. Now Jaworski quickly hired Chilton as

a consultant, and for good measure he also signed up Jeff
Schell. From then on Monsanto would have immediate access
to all the cutting-edge plant genetic engineering work going
on in the world. This would now be a three-way race:
between the public sector university-based work of Van
Montagu and Schell, Chilton, and Monsanto, the new
corporate entrants from St Louis.

Ernest Jaworski had by 1979 long been pushing to get his
bosses in Monsanto to take the emerging field of plant
biotechnology seriously. Monsanto was historically a
chemicals business, and successful Monsanto executives were
chemicals people who had built their careers on selling
pesticides and plastics. Jaworski was interested in biology
more than chemistry. He saw that the world was changing,
and that pesticides were becoming controversial because of
their negative impacts on human health and the wider
environment. As early as 1972 Jaworski was asking his
Monsanto superiors to create a cell biology research
programme that would hopefully yield a new generation of
crop-protection products not involving the widespread use of
chemical sprays. As he put it in a 1996 interview for the
internal Monsanto magazine, 'I started thinking about what
would happen to Monsanto in the future. After you've
invented all the herbicides you need, all the insecticides you
need, all the fungicides, what are you going to do to keep
growing? I concluded that a time would come when you
couldn't solve all problems with chemicals.'[5]

Watching Chilton and Van Montagu's work, Jaworski
knew that time was running out if Monsanto was to become
the first major commercial player in the nascent field of
plant genetic engineering. As his bosses prevaricated, both
Chilton and Van Montagu's results kept coming. Within a
couple of years both teams had published papers confirming
the exact location in the plant cell of *Agrobacterium*'s T-DNA
(it was in the nucleus, as expected). Jaworski was impatient

because he knew that simply keeping up to date with Chilton and Van Montagu's research was not enough. Monsanto would have to replicate and improve on this work in-house. If Monsanto were to have a future 'proprietary position' (as one internal company memo put it[6]) in this new field it would need to hold patents. That in turn meant it would need to make its own inventions. The timing was expeditious as the path towards full commercialisation of biotechnology had just been cleared by the US Supreme Court, which ruled in the landmark 1980 *Diamond v. Chakrabarty* case that novel microorganisms created in the laboratory could be protected as intellectual property under US patent law.[7]

Jaworski set out to hire research scientists who could begin to lead the genetic engineering field from within the company. One of these was Robert Fraley, a young post-doctoral fellow at the University of California San Francisco (UCSF), and already a pioneer in the field of plant DNA transfer. Fraley, smart and ambitious in equal measure, was interested in corporate management as well as lab work. Today he is Executive Vice President and Chief Technology Officer of the Monsanto company, second in influence only to the CEO, Hugh Grant. Fraley more than anyone shared Jaworski's vision of plant genetic engineering as a stepping stone for Monsanto to move from the age of chemistry to the age of biology in crop protection. It was his job to turn this vision into reality – a reality that not only would transform agriculture, but would hopefully open up whole new markets and make billions for Monsanto in the process.

'As a kid I always knew I wanted to be a scientist,' Robb Fraley told me when I reached him by phone for an interview. Fraley grew up on a farm in Illinois, 100 miles south of Chicago. 'We had about 300 or 400 acres of land, raised the normal range of Midwest crops: soybeans and corn and wheat, and a little bit of livestock,' Fraley recalled.[8] But speaking to him I got the feeling that the young Robb decided very early on that farming was not for him. 'I always tell folks I was one of those really strange kids who always knew I would be a scientist. I can remember when I was five or six years old copying pictures out of encyclopedias, and sneaking into my dad's or my grandpa's workshop and trying to take things apart and figure out how they worked. I always had that feeling that I would do something in science or technology,' he remembered.[9] Fraley was the first in his family to go to college: 'Farming was good, but what I realised early on was that my dad was a really small farmer, we were a pretty poor farm family.' When his father died 'pretty young' that was the end of any agricultural ambitions Fraley might have entertained. After an undergraduate degree at the University of Illinois, Fraley went on to do postgraduate work at the UCSF. Looking around from UCSF for job options, Fraley connected with Monsanto 'a bit randomly', as he put it. Attending an academic conference in 1979, he met a Monsanto scientist who told him, 'You know, Monsanto is getting ready to build an ag biotech program, and before you do anything else, you should talk to this guy Ernie Jaworski who's going to set it up.' It turned out that Jaworski was passing by to attend another conference, and the two men met between planes at Boston airport. According to Fraley, 'Ernie was a very charming and passionate leader, who quickly was able to convince me that I needed to at least come to St Louis and talk to them about what Monsanto was doing. And that was the seminal connection. A few months later I visited St Louis. I met with Ernie, listened to what he was trying to put together, talked about the team ... and I guess 37 years later ... the rest is history.'

The proposed *Agrobacterium* genetic engineering system presented some hefty challenges to the stellar new Monsanto team. Their first problem was to find a 'selectable marker' for transformed plant cells. In other words, there had to be a way to screen the thousands of cells that had been subjected to the *Agrobacterium* treatment in order to find the ones that had been successfully 'transformed' with the desired new DNA sequences (most of the time, for one reason or another, it didn't work). Their solution was to add a gene for resistance to the antibiotic kanamycin into the *Agrobacterium* plasmid, so that successfully transformed plant cells would be able to grow in a Petri dish containing this otherwise toxic compound, while all the non-transformed cells would die. At the same time, the tumour-causing genes in the *Agrobacterium* plasmid had to be knocked out: Monsanto was interested in growing healthy crops, not cornfields full of misshapen crown galls. Monsanto's new biotech team also had to face down scepticism from both inside and outside the company. 'There were scientists at the time who believed that plant DNA was so different that you couldn't put a gene in,' Fraley recalled. 'There was the belief that these genes would not express in plant cells, or that the particular cell that you would get the gene into would be unique and would not be regenerable.'

Like Chilton before him, Robb Fraley was soon to have his own moment of epiphany. As he related, 'I remember Rob Horsch running – we had a long hallway – down the center of U Building, and he was just screaming, "It worked! It worked! It worked!" Now Rob's a short guy, but he was jumping up and down and bouncing off the ceiling. The results were absolutely clear. You could look at both Petri plates, and the ones with the gene had lots of growing green colonies, and the ones without the gene were all dead and brown. It worked and it was so clear. And that was a huge moment.' As Fraley told me: 'It was so exciting because in one fell swoop we realised we'd solved the problem.' Only a few weeks later Jeff Schell was over from Europe visiting the Monsanto lab. As Fraley remembered, 'I can still remember showing Jeff Schell the first time that we had the definitive proof that we could introduce a gene into plant cells. We were looking together at the Petri plates and Jeff started crying. And I said, "Jeff, what's wrong?" And he said, "You've done it! You've done it!" And that was when what we had accomplished hit me.' The final

hurdle to full-scale plant genetic engineering had finally
been cleared.

The true historical moment, what has been called the
'symbolic coming-of-age of plant genetic engineering', took
place a couple of months later on 18 January 1983. At the
Miami Biochemistry Winter Symposium all three teams,
represented by Mary-Dell Chilton, Jeff Schell and Rob
Horsch, made back-to-back presentations revealing similar
breakthroughs. And science was no longer the only game in
town; the lawyers too had been busy. Monsanto had filed for
a patent only the day before the Miami meeting, on 17
January. But Van Montagu and Schell had already filed their
own patent application in Europe the day before Monsanto.
Within months, Chilton would have her own corporate
affiliation, jumping ship from Washington University in St
Louis to Ciba-Geigy, today merged into the Swiss
agrochemicals giant Syngenta. The Ghent team declined to
join a big corporation. Instead they established a startup
company to take on the *Agrobacterium* patent, Plant Genetic
Systems. Like many startups, it didn't remain independent for
long. Within a few years it was snapped up by Hoechst, now
part of the German seeds, pharmaceuticals and chemicals
giant Bayer. Schell was by then also having a successful career
as director of the Max Planck Institute for Plant Breeding in
Cologne, Germany, responsible for a team of well over a
hundred research scientists.

The Miami symposium episode has been seen by some as
the moment where Monsanto stole the limelight, muscling
aside the smaller players. In his 2001 book *Lords of the Harvest*,
the National Public Radio food journalist Dan Charles quotes
Van Montagu as complaining that he was 'frustrated because
Monsanto monopolised the media' and that 'no one listened
to our story'. Charles reported, 'Horsch was accompanied to
Miami by a representative from the company's public relations
department. Monsanto had sent out a press release. Within

days, the front page of the *Wall Street Journal* credited Monsanto with a scientific breakthrough.'[10]

However, the passage of time seems to have smoothed any ruffled feathers. I got no trace of resentment from my own interviews with both Van Montagu and Chilton. Speaking about Monsanto, Van Montagu told me that 'I always considered them highly', though he conceded that the external friendly face the company's representatives portrayed may have been different from the internal reality. He also 'absolutely' credited Ernest Jaworski for having the vision of how to commercialise plant genetic engineering, and pushing Monsanto into embracing it. Van Montagu also told me that Jeff Schell and Jaworski 'were really close personal friends', so in his view the relationship was unlikely to have been economically exploitative.

The *Wall Street Journal* story of 20 January 1983 referred to by Dan Charles was titled 'Monsanto Scientists Say They Succeeded In Inserting Foreign Gene Into Plant Cells'. In the second paragaph the reporter noted: 'The St Louis-based chemical producer said scientists in Belgium accomplished a similar feat about the same time, working independently of the St Louis team ... The Monsanto experiment was carried out by Robert Horsch, Stephen G. Rogers and Robert T. Fraley. They credited Prof. Schell of Belgium and Mary-Dell Chilton of Washington University in St Louis for the work with the *Agrobacterium* that made the experiment possible. The experiment in Belgium was reported by Prof. Schell and Marc Van Montagu.'

Getting proper credit for scientific discoveries is important for more than just egotistical reasons. It has serious legal and financial implications, as the battling patent applications showed. So were Monsanto the first bio-pirates, stealing others' inventions? Mary-Dell Chilton was quoted in *Lords of the Harvest* as saying that her emotions towards Monsanto were 'some terrible mixture of jealousy and respect and admiration and anger', but when I spoke to her a decade and a half later Chilton said she didn't have any strong feelings. 'No, I think that the interaction was fairly balanced,' she added, after a thoughtful pause. That balance may have tilted a little in Monsanto's direction, however: 'I think Monsanto probably got more than their half out of it, because what Monsanto had was not unique, and what I had was unique. Monsanto had money, and I had knowledge and technology and experience.'

She conceded that Monsanto 'solved that problem' by hiring both her and Jeff Schell as consultants. 'So you couldn't really complain then when Monsanto ran off with the patent rights,' I suggested in our phone interview. There was another thoughtful pause. 'Well, it depends. If Monsanto took an idea of mine and said it was their vision, that probably was not true. Under United States patent law the inventor is the person who has the idea. In some cases perhaps I should have been named as one of the inventors. And they didn't do that. They never did that. It would have muddied their legal waters to name anybody outside Monsanto as a co-inventor.' She interrupted herself: 'But it's not important. In my mind it's not important.'

Thirty years after their discovery of plant genetic engineering, surviving representatives from the competing three teams were once again united on the evening of 16 October 2013. In a grand room in Des Moines, several hundred hushed assembled dignitaries, all smartly dressed as appropriate to this highly formal occasion, and seated in semi-circular rows in the chamber of the magnificent Iowa State Capitol building, awaited their entrance as 2013 World Food Prize Laureates. A long sequence of presidents, prime ministers, ambassadors, senators and state governors all processed in to take their seats, as an unseen announcer's voice boomed out to announce each in turn. Then six trumpets rang out in a spectacular fanfare, and the announcer said: 'And now, welcome the World Food Prize Laureate party, and distinguished guests of honour – the 2013 Laureates, Marc Van Montagu of Belgium! Mary-Dell Chilton and Robert Fraley of the United States!'

On the right was Van Montagu, in a suit with a purple tie. In the middle, walking with a stick, but still with a regal bearing as befitted the woman once dubbed the 'Queen of *Agrobacterium*', was Chilton. On the left, in a navy suit with a sky-blue tie, Monsanto's Robb Fraley. Looking straight ahead, the three slowly descended the steps into the packed auditorium. Facing them, on the polished wooden dais in front of the row of trumpeters, sat the World Food Prize sculpture, an earth-coloured stone bowl made of alabaster, with a pewter sphere sitting in its centre.[11] There was also a substantial cash prize. The whole occasion was expressly designed at its foundation in 1990 to resemble the Nobel Prize ceremony, at the behest of its founder Norman Borlaug, father of the Green Revolution.

On giant video screens the image flashed up of M. S. Swaminathan, the Indian crop breeder who worked with Borlaug to bring new high-yielding varieties of wheat and rice to the Indian sub-continent. 'I am particularly happy that on the sixtieth anniversary of the discovery of the double helix structure of the DNA molecule, three very eminent biotech-nologists – Professor Marc Van Montagu, Mary-Dell Chilton and Robert Fraley – have been recognised for their work and they will be receiving the World Food Prize,' Swaminathan said in his video address. 'It is very appropriate because I think the science of genetic engineering, the new biology and new genetics, has certainly opened up completely new opportunities.'

The video cut to the *Agrobacterium* story, with old photos of Marc Van Montagu in 1960s garb of thick glasses, mop-top black hair and a patterned shirt, together with the bearded Jeff Schell. There followed a graphic of *Agrobacterium* delivering its genes into the plant cell, and pictures of plant roots disfigured by the bacterium's crown gall tumours. Then the video cut to childhood photos of Mary-Dell Chilton, with her boyish fringe and determined scowl. Then came a photo of her ordinary-looking high school in Illinois, and another of Chilton as a young mother with a baby in a backpack, photographing a plant in a field. The video history shifted to Chilton's 'brute force' experiment to discover the *Agrobacterium*'s spliced DNA inside the plant cell.

Then the 74-year-old Chilton was speaking pre-recorded from the screen, wearing bifocals and a pink lab coat, with shelves full of bottles of chemicals and other laboratory equipment behind her. 'In a real sense, the process that we used for genetically engineering a crop plant is a natural one,' she said, echoing Jeff Schell from 15 years earlier. 'We learned it from nature. We learned it from *Agrobacterium*, a little bacterium that did this before we ever discovered it. All we did was learn how *Agrobacterium* manages to put a gene into a plant, and we copied that process. We exploited that natural process to put genes into plant cells, genes of our choice to benefit the farmer and the end user of that plant.'

Then the video profile moved to Robb Fraley. There was Fraley as a child, sitting in a toy tractor on the family farm outside a typical mid western white clapboard house. Then an older Robb on a real tractor, then at Monsanto, on the back of a planter, sitting next to his colleagues Rob Horsch and Steve Rogers. Then there were the bespectacled Horsch and Rogers holding up their Petri dishes of blobby green plant calluses – the first successfully transformed cells, ready for growing out into new genetically engineered plants. The video graphic credited the Monsanto team with the discovery of how to delete the tumour-inducing genes from *Agrobacterium*'s plasmid, and replace these with recombinant DNA consisting of desired new genes ready for insertion into the target plant. There stood Fraley with a bottle-brush moustache, smiling in front of some tomato and petunia plants, then an older Fraley in suit and tie on a platform behind the Monsanto logo, a US flag in the background. Fraley's screen interview was shorter, but characteristically upbeat: 'I feel really privileged, and I think for me it's just the beginning of a wave of innovation that's going to be so important for agriculture.' The video ended with scenes of modern agriculture around the world, backed by uplifting music and a breathless voiceover extolling the virtues of the plant biotechnology revolution that Van Montagu, Chilton and Fraley as 2013 World Food Prize Laureates had helped to pioneer.

Sadly absent was Jeff Schell, who had died a decade earlier. Schell's 1998 Japan Prize-winning address perhaps expresses most elegantly the vision of the early pioneers that genetic engineering could become a tool for a more sustainable agriculture. 'Agriculture, as it is practised currently, is one of the biggest sources of environmental pollution,' Schell noted. 'If one wants to diminish the negative impact of agriculture on the environment, one should optimize productivity, i.e. maximum yield and quality for a given input ... Plant breeding is one of the few, and one of the most effective methods, to improve agricultural productivity without simultaneously destroying the environment. This is true for the industrialized world, perhaps even more true for the developing world and holds for both intensive and extensive agriculture. If plant breeding is to contribute to the solution of the enormous problems which we must face in the next decades, then the best techniques must be used including genetic engineering.' However, by 1998 Schell was of course well aware that 'new science and technology are viewed with caution and indeed fear', this fear being 'particularly pronounced in Europe'. For Schell, this posed a conundrum: 'Unfortunately, precisely the organizations and political parties whose support is based on environmental protection ... have been most active in rejecting plant biotechnology. The potential of this new technology to protect the environment has largely been ignored.'

As Van Montagu asked plaintively in a retrospective article written decades after his initial scientific discoveries: 'We live in a world where more than a billion people are hungry or starving, while the last areas of tropical forest and wild nature are disappearing. Why could this new technology not bring the solutions to these challenges? Why has this not happened yet? What did we do wrong?' In my view, the snowballing of opposition to plant genetic engineering resulted not so much from decisions taken by Schell, Van Montagu or even Chilton, but largely because of the dominating presence of their erstwhile competitor, Monsanto.

CHAPTER FOUR

A True History of Monsanto

It was mid-October 1901. In the booming, smoke-wreathed, sprawling city of St Louis, Missouri, the so-called 'Gateway to the West', a fearsome-looking middle-aged man with dark red hair and a sandy moustache was pacing the downtown area. Just a block from the Mississippi river, John Francis Queeny was seeking a location for his planned new business venture. Born in industrial Chicago in 1859, the eldest of five children in a hard-working but poor second-generation Irish immigrant family, Queeny had just $1,500 to his name. Having borrowed another $3,500 from an associate with which to launch a chemicals firm, he was about to make a very risky bet.

Queeny had already worked his way up from the streets, leaving school at the age of 12 to take a job as a barrow boy with a pay packet of $2.50 a week. By 1894 he had risen to become a sales manager of the pharmaceutical company Merck & Co., and had now spotted an opportunity to strike out on his own. Even so, he planned to keep the Merck job going as backup, and his boss had told him not to use the Queeny name in the new operation to avoid confusing customers. So John Francis decided instead to use the less well-known maiden name of his new wife, whom he had married five years earlier in Hoboken, New Jersey. With a Spanish father and an aristocratic pedigree of sorts, she had a very different background from her hard-bitten Irish husband. Her name was Miss Olga Mendez Monsanto.[1]

John F. Queeny's business idea was a simple one. Knowing that a German company had a monopoly over the manufacture of the recently discovered super-sweet chemical saccharin, Queeny wanted to establish a US base for saccharin manufacture to challenge German imports with home-made sweetener. His reasoning was logical enough. Saccharine already had a big market in candy, soft drinks and chewing tobacco, and 1lb could pack the same sweetness as 300lb of real sugar.[2] Even accounting for the fact that saccharine was more expensive to make, in terms of sweetening ability it was six times cheaper pound for pound than cane sugar. Saccharine did have the drawback of a strange aftertaste, an uncertain safety profile and the somewhat unappealing fact that it had been accidentally discovered as a by-product of coal tar. But Queeny, with his pharmaceuticals background, was undaunted by these sorts of concerns. His instinct was good: artificial sweetener was to become one of the totemic products of the chemicals age.

As his first employee, Queeny hired a newly qualified Swiss-German chemist called Louis Veillon, who already knew the complex recipe for saccharin manufacture, having learned it back home in Europe. Veillon's first task was to assemble the necessary equipment, second-hand where possible to save money. This included an old steam engine, an even older boiler, pump, scales, kettle, pipes and a brand new centrifuge, all installed at the newly established Monsanto Chemical Works headquarters on South Second Street, a dingy location next door to a match factory. (The adjacent building was, predictably enough, destined to burn down spectacularly a few years later. Queeny, ever the opportunist, bought out the wrecked neighbouring plot and expanded.)

The first batch of saccharine was synthesised in February 1902, although – so the story goes – both Veillon and Queeny's taste buds were so deadened from saccharine fumes inhaled in the factory that they initially could not taste any sweetness at all. Thinking that the whole venture had been a failure they decamped in misery to a nearby restaurant, where a waiter commented 'My, that's sweet!' when offered a sample. Celebrations duly ensued, and Veillon received his first bonus from his grateful employer: a box of Havana cigars. Queeny sold his saccharine to a soft drinks manufacturer, adding caffeine and synthetic vanilla flavouring to his roster a few years later in order to serve the same rapidly growing soda market.

Although the new Monsanto company had some hard times ahead – at one point Queeny was reduced to selling his own horse and buggy to get through a price war launched by German saccharine manufacturers – by 1915 it was turning over a million dollars, and 50 years later Monsanto would break the billion-dollar sales barrier. Under the leadership of John F. Queeny's son Edgar the company enjoyed rapid twentieth-century growth on the back of pharmaceuticals (it was the largest manufacturer of generic aspirin from 1917 onwards), plastics and numerous other chemical products. Some of the ingredients for the artificial rubber in the tyres of the jeeps on which the US Army rolled through the Pacific and Europe in the Second World War came from Monsanto factories. Monsanto chemistry even contributed to the top-secret Manhattan Project, helping in the production of plutonium for the nuclear bomb that destroyed Nagasaki. By the 1950s, the company was churning out everything from nylon fibres to automobile interiors in ever-growing quantities.[3] Monsanto even invented artificial turf in 1965: originally called 'Chemgrass', it was rebranded as AstroTurf after it was installed at the Houston Astrodome, a multi-sports domed stadium, in 1966.[4]

Monsanto had a fairly typical mid-twentieth-century vision for how modern technology could reshape American lives for the better. An all-plastic 'Monsanto House of the Future', set in front of a small-scale concrete replica of the Matterhorn, was a popular Disneyland attraction during the decade after 1957.[5] This was the dawn of the Space Age, when human ingenuity and progress apparently knew no limits. The era's boundless technological optimism became tarnished during the 1960s, however, thanks to the traumatic American experience in Vietnam and a growing awareness of the environmental downsides of rapid industrial growth. The Monsanto House of the Future was dismantled in 1967, by which time the alienating artificiality of an all-plastic living space no longer represented the kind of future that the new generation aspired to. Scrapping the Disneyland structure was far from easy, however. The wrecking ball simply bounced off the plastic exterior, so the demolition crew had to come back with hacksaws and blowtorches and spend two weeks taking the Monsanto House of the Future apart piece by laborious piece.

Of course, Monsanto was never going to have much in common with the Summer of Love hippies, who were more interested in reconnecting with nature (and each other) than trusting to Monsanto's vision of a high-tech world constructed of plastic and chemicals. This was especially the case because Monsanto was one of several companies contracted by the United States government to produce the herbicide 2,4-D (2,4-Dichlorophenoxyacetic acid), which when combined with 2,4,5-T (2,4,5-Trichlorophenoxyacetic acid) formed a powerful defoliant or weedkiller. This particular two-way combination was not intended for use on US farms, however, unlike the other herbicides that had built up big markets in the post-war US corn belt. Instead, the US Department of Defense began shipping huge quantities of this weedkiller to help the war effort in Vietnam. It was transported in large drums with orange stripes on the side, for which it later became infamous worldwide by its code name, Agent Orange.

Agent Orange devastated the Vietnamese environment, as it was expressly intended to do. The US National Academy of Sciences (NAS), in a 2012 report entitled 'Veterans and Agent Orange', stated: 'The herbicides were used to defoliate inland hardwood forests, coastal mangrove forests, cultivated lands, and zones around military bases.' Between the commencement of large-scale spraying in August 1965 and its cessation amid increasing controversy in February 1971, some 18 million gallons (69 million litres) of herbicide were sprayed over roughly 3.6 million acres (1.5 million hectares) of North and

South Vietnam.[6] About 5 per cent of the entire country was defoliated with this toxic weedkiller.

Contrary to popular myth, the toxicity of Agent Orange did not come directly from the herbicides themselves; 2,4-D is still used widely in agriculture, and is not generally considered a carcinogen in either humans or animals.[7] The problem came with 2,4,5-T, which as a by-product of its manufacturing process was contaminated with TCDD, identified by the NAS as 'the most toxic form of dioxin'.[8] The NAS publishes regular updates to its list of health impacts thought to be associated with Agent Orange. Its 2012 report put soft-tissue sarcoma (heart), lymphoma, chloracne (a serious skin rash) and some types of leukaemia in its 'Sufficient Evidence of an Association' category, with a much longer list including other cancers, Parkinson's disease, stroke and spina bifida in offspring of exposed people in the 'Limited Evidence' category. There remains an even longer list in the 'Insufficient Evidence' bracket, illustrating the continuing controversy and scientific uncertainty about the precise scale of Agent Orange health damages in exposed populations.

Monsanto was not the only corporation responsible for manufacturing Agent Orange and selling it to the US Department of Defense. A class-action lawsuit launched by US Vietnam veterans in the 1980s named 19 companies as defendants with the Dow Chemical Company, the Monsanto Company, the Diamond Shamrock Corporation, Hercules Inc. and the Thompson Hayward Chemical Company as principal defendants. The subsequent years of litigation and argument were largely fronted by Dow, not Monsanto. In March 1983 Dow's president, Paul Oreffice, maintained that dioxins were not toxic enough to cause anything more than a rash, a position that the company's lawyers doggedly stuck to in court for years despite increasingly compelling evidence to the contrary. 'There is absolutely no evidence of dioxin doing any damage to humans except for something called chloracne,' Oreffice insisted when interviewed on NBC's *Today* show at the time. Those exposed to dioxins in Agent

Orange should stop worrying, Oreffice maintained, because Dow's own studies showed that 'there is no evidence of any damage other than this rash which went away soon after'.[9]

However, Dow's president should have known even then that he was on thin ice. Just a month after his insistence that dioxin caused little more than teenage spots, internal company memos surfaced showing that as far back as 1965 Dow was aware of the toxic properties of dioxins in exposed laboratory animals. Indeed, the company was sufficiently alarmed as to invite representatives of competing chemicals manufacturers to a confidential meeting (Monsanto was invited but did not attend, according to later reports in the *New York Times*) where industry scientists outlined the dangers. Dow's concern, as later revealed in memos written by those attending the secret meeting and published in the *New York Times*, was not so much about protecting public health as that news leaking out might make the situation 'explode' and invite the unwelcome prospect of federal regulation.[10] 'They [Dow] are particularly fearful of a Congressional investigation and excessive restrictive legislation on the manufacture of pesticides which might result,' wrote one of those present at the secret briefing. For its part, Monsanto insisted that their company 'didn't do any testing, period, not then and not now'.

The question of whether or not the herbicide's manufacturers knew Agent Orange was potentially toxic was critical in their legal defence against the class-action lawsuit by Vietnam veterans. Dow and Monsanto's lawyers claimed 'military contractor defense'. In other words, the decision – and resulting liability – for the use of Agent Orange in Vietnam rested with the Pentagon, they argued, not the companies, who merely supplied it on contract. Since the US Government enjoys sovereign immunity from legal action, the Vietnam veterans would then have no recourse to damages. But if the companies knew about the dangers of dioxins and didn't tell the government, at least until some years later, the manufacturers might themselves be legally liable as a result of this alleged cover-up.

The issue was additionally sensitive because dioxin pollution was not just restricted to war-torn south-east Asia. The 1978 Love Canal scandal related to dioxin contamination in upper New York state. Dow fought the Agent Orange case hardest in court because it probably had most to lose: at a price of $7 a barrel, it had supplied fully a third of the government-contracted Agent Orange formula used in Vietnam, more than the other defendants including Monsanto.[11] Following six years of tortuous arguments the companies settled the case on the eve of the trial in 1984, contributing $180 million to a damages fund for veteran claimants and their families. In 1993 the *New York Times* revealed that Monsanto 'bore the brunt' of the 1984 settlement, forking out 45.5 per cent as compared to Dow's 19.5 per cent; the reason why is not explained.[12]

In 2005 a lawsuit on behalf of Vietnamese civilians alleging that the US was guilty of war crimes in Vietnam was dismissed by a federal judge, who ruled that the Geneva Convention prohibition on chemical weapons 'extended only to gases deployed for their asphyxiating or toxic effects on man, not to herbicides designed to affect plants that may have unintended harmful side effects on people'. Scot Wheeler, a spokesman for the Dow Chemical Company, was unrepentant. 'We believe the defoliant saved lives by protecting allied forces from enemy ambush and did not create adverse health effects,' he said. A spokesman for Monsanto complained with unintended irony that 'there's so much emotion wrapped up in cases like this'.[13]

As the tide of public opinion began to turn against them in the late 1960s and early 1970s, Monsanto and the other chemical companies fought a bitter rearguard action to protect their markets and reputations. Undoubtedly the most severe blow came with the publication of Rachel Carson's book *Silent Spring*. Carson issued strong warnings about the damage done to birds and other wildlife by the overuse of agricultural chemicals in US farming, DDT in particular. The enduring importance of her book can hardly be overstated. Many would date the origin of the modern environmental movement to the morning of 27 September 1962 when *Silent Spring* began to roll off the presses.

Carson was a true pioneer. Not only was she the first to bring the direct biological damage from pesticides to the wider attention of society, she was also ahead of her time in the way she wrote about the complex dynamics and interrelatedness of ecosystems. Her earlier books on the oceans, also bestsellers, were as poetic as they were scientific, for she was a gifted writer. This quality made her broadside against chemical agriculture all the more devastating. She warned in *Silent Spring*: 'These sprays, dusts and aerosols are now applied almost universally to farms, gardens, forests and homes, non-selective chemicals that have the power to kill every insect, the good and the bad, to steal the song of birds, the leaping of fish in the streams, to coat the leaves with a deadly film, and to linger in the soil – all this, though the intended target may be only a few weeds or insects. Can anyone believe it is possible to lay down such a barrage of poisons on the surface of the earth without making it unfit for all life?'[14]

Carson's critique went deeper than just the misuse of chemicals in farming and disease control. She also voiced her feeling that 1950s American society was putting too much faith in technological progress as the cure for all ills, writing about the modern era as a 'Neanderthal age of biology and philosophy, when it was supposed that nature exists for the convenience of man'. She worried that many of the 'new chemical and physical agents' of the industrial age might also be carcinogens for which humans had evolved no protection. The subject of carcinogenicity had personal relevance for Carson, who was already ill with cancer when the book was published. Invited to testify before a June 1963 Senate subcommittee on pesticides, she was barely able to walk to her seat at the panel table. Carson had already survived a radical mastectomy and wore a brown wig to hide the baldness that was the result of her ongoing treatment.[15] Sapped of energy, she was too unwell to conduct much publicity for her book, and died in January 1964 at the age of

56, unaware of the transformative impact her work would have on the world.

Carson's own physical vulnerability served to highlight even more the ugliness of the chemical industry's attacks on her. Its chief spokesman was Dr Robert White-Stevens, who appeared on a contemporary CBS report in spotless lab coat, surrounded by laboratory equipment. Speaking in clipped tones, he asserted: 'The major claims in Miss Rachel Carson's book *Silent Spring* are gross distortions of the actual facts, completely unsupported by scientific experimental evidence and general practical experience in the field.'[16] In fact Carson had been extremely careful in writing the book, taking four years over the project and working closely with many scientists in gathering her evidence. White-Stevens continued: 'If man were to faithfully follow the teachings of Miss Carson, we would return to the dark ages, and the insects and diseases and vermin would once again inherit the Earth.'

Behind the scenes, the industry did its best to stop *Silent Spring* even coming out. DDT manufacturer Velsicol threatened to sue both the publisher Houghton Mifflin and the *New Yorker* magazine, which serialised several chapters of Carson's work. The company insinuated in a threatening legal letter to Houghton Mifflin that Carson was probably a Communist – a serious matter in those immediate post-McCarthy years. If Carson's work led to the elimination of pesticides, Velsicol charged, 'our supply of food will be reduced to East-curtain parity'. Others tried to dismiss Carson as just a typically hysterical female. 'Isn't it just like a woman to be scared to death of a few little bugs?' one correspondent sneered in a letter to the *New Yorker*. The National Agricultural Chemicals Association funded a countrywide public relations campaign, buying newspaper adverts, firing off letters to the editor and publishing pamphlets, all aimed at reassuring an increasingly worried public that agricultural pesticides were nothing to be concerned about.[17]

Monsanto also went on the attack. In the October 1962 edition of the *Monsanto Magazine* the company carried an unusual article headlined 'The Desolate Year', an unsubtle parody of the opening chapter of Carson's book, which had been entitled 'A Fable for Tomorrow'. While Carson had written about an imaginary American town where the wildlife and domestic animals had begun to die due to pesticide poisoning, Monsanto's effort used similarly lyrical language to portray an imaginary year during which the entire nation was deprived of the benefits of insecticides.

'For now spring came to America – an extremely lively spring,' the anonymous author wrote, alluding to the 'silent spring' of Carson's title. 'Genus by genus, species by species, sub-species by sub-species, the insects emerged. Creeping and flying and crawling into the open, beginning in the southern tier of states and progressing northward. They were chewers, and piercer-suckers, spongers, siphoners and chewer-lappers, and all their vast progeny were chewers – rasping, sawing, biting maggots and worms and caterpillars. Some could sting, some could poison, many could kill.'[18] As the apocalyptic tide of insects swarmed forth, the article continued, 'the garrotte of Nature rampant began to tighten'. Monsanto was clear about the solution: 'Pesticides are essential to maintain and improve our food supplies and public health.' The magazine article was copied and sent out to newspapers around the country, accompanied by a five-page 'factsheet' about the benefits of agricultural chemicals.

The fierceness of the industry's assault was belied by the rather moderate nature of Rachel Carson's own views. She did not argue for total elimination of pesticides, even DDT, acknowledging that insect populations needed to be managed for successful food production and disease control. Nor did she argue – as implied in the *Monsanto Magazine* rejoinder – that insect pests should be left to run rampant across America's defenceless cornfields. One of her strongest arguments against the overuse of pesticides was that their very usefulness was being wasted because of the rapid evolution of resistance in insect pest populations. 'Practical advice should be "Spray as little as you possibly can", rather than "Spray to the limits of your capacity'", she quoted one expert as saying about malaria control in the penultimate chapter of *Silent Spring*. And biological pest controls using viruses and pheromones offered more environmentally benign but equally effective alternatives, she suggested. Carson's view that indiscriminate use of insecticides was killing beneficial and predator insects and thereby harming overall pest control has today become received wisdom among agronomists.

In any case, powerful people were by now listening more to Carson than to the chemicals manufacturers. Concern went right to the top: President John F. Kennedy appointed a commission to investigate Carson's claims which met for a year and ended up largely vindicating her work. DDT was banned for domestic US use in 1972, and rising public interest in environmental issues led to the creation of the federal Environmental Protection Agency as a major department of government during the Nixon administration. Contrary to the assertions of many anti-green campaigners, who blame Carson and the environmental movement for supposedly causing millions of deaths from malaria after the banning of DDT, the insecticide continued to be used in Asia and Africa for decades afterwards. As late as 2007, 3,950 tonnes of DDT were sprayed in developing countries, according to the United Nations.[19]

Monsanto was stung not only by the Agent Orange scandal and Rachel Carson's epic assault on pesticides, but also by later public condemnation for its role as the main US manufacturer of PCBs. These inert flame-retardant, heat-conducting substances were initially thought of as 'miracle chemicals' and used in everything from electrical equipment to newsprint and frying pans in the 1950s and 1960s. Monsanto ceased production in 1977, but by that time abundant evidence had accumulated about both the environmental persistence and the carcinogenicity of PCBs, and of how the company had dumped large quantities in waterways and landfills.[20]

Today Monsanto remains embroiled in legal actions as a result of this legacy.[21]

Monsanto fought back against declining public trust with an advertising campaign using the banner 'Without chemicals, life itself would be impossible'. One ad appearing in *National Geographic* in 1977 showed a blonde child nose to nose with a cute family dog in an idyllic-looking sunlit pasture. 'Some people think anything "chemical" is bad and anything "natural" is good. Yet nature *is* chemical,' the print below explained. It went on to describe the 'chemical process called photosynthesis' and how 'a chemical called vitamin D' is important to avoid rickets. Then it got to the nub of the argument: 'Chemicals help you eat better. Chemical weed-killers have dramatically increased the supply and availability of our food. But no chemical is totally safe, all the time, everywhere. In nature or the laboratory. The real challenge is to use chemicals properly. To help make life a lot more liveable.'[22]

But Monsanto was on to a loser if it thought it could reclaim the word 'chemical' from rising public distrust and recast it as something good. Agent Orange and PCBs were just two scandals among many, and Monsanto was only one of a number of chemicals companies embroiled in them. Dow also supplied the US military with napalm between 1965 and 1969, which had even more horrific immediate impacts on Vietnamese civilian populations than Agent Orange. In 1984 the world's worst industrial accident took place in the Indian city of Bhopal, killing thousands of people as toxic gas leaked out of a badly run pesticide factory.[23] The Bhopal plant was owned by Union Carbide, whose chief executive Warren Anderson never faced trial in India or anywhere else, despite an international campaign to hold him accountable. Anderson died in comfortable retirement at Vero Beach in Florida in 2014 at the age of 92.[24]

Some chemicals companies had even deeper and darker histories that continued to weigh on their modern reputations. BASF and Bayer were successors to the infamous IG Farben, for example. This was the German company whose subsidiary manufactured Zyklon B gas pellets, used by Hitler to exterminate millions of concentration camp victims between 1942 and 1944. IG Farben also ran a slave labour plant conveniently located right alongside Auschwitz.[25] Although IG Farben was split up after the war and its key leaders put on trial for crimes against humanity, all were released early and many went on to enjoy successful careers in the various successor companies whose brands are still familiar today.

It is far from obvious why among all this strong competition Monsanto should have become singled out as 'the most evil corporation in the world', as one so often hears today. Every year a global 'March Against Monsanto' brings together activists who loathe and detest everything they think the company stands for. Monsanto consistently makes the top five in annual lists of 'most hated companies'.[26] When the owner of one start-up 'big data' company sold his stake to Monsanto, his own father was aghast, responding 'Monsanto? The most evil company in the world?'[27] Internet conspiracy theories abound. One popular myth is that Monsanto refuses to serve GMOs in its own cafeteria; another is that it took over and now runs the infamous mercenary firm Blackwater, whose employees were implicated in human rights abuses in US-occupied Iraq.

Not surprisingly this high degree of public distaste affects the company's staffers on a daily basis. I was once told a story by a mid-level Monsanto employee, a sales agent in a US farming state. This man had made the mistake of wearing a shirt that had a small 'Monsanto' logo on it while travelling on a plane. The air hostess told him that she would refuse to serve him during the flight because of the 'terrible things your company does'. The bewildered sales agent told me with a self-deprecating laugh, 'I'm a middle-aged white guy; I'm not used to being discriminated against!'

My conclusion is that it was the later development of GMOs that catapulted Monsanto into the forefront of public vilification, not its earlier activities as a chemicals manufacturer. This is somewhat ironic, because its move into genetic engineering was probably the most environmentally friendly thing that Monsanto ever did. Certainly the company's early interest in biotechnology was stimulated precisely because it wanted to reduce its dependence on pesticides and other chemicals in response to Rachel Carson's warnings.

Monsanto accordingly made enormous investments in biotech. In 1979 the then-CEO John Hanley hired Howard Schneiderman, dean of the School of Biological Sciences at the University of California, Irvine, as head of research, with a special focus on the then rapidly emerging field of biotechnology. Schneiderman started out by spending $165 million constructing and equipping a new Life Sciences Research Center on a 210-acre campus outside St Louis, Missouri. As the *New York Times* put it in a 1990 retrospective, 'In the center's four buildings and 250 laboratories, 900 researchers spend their days hunched over Petri dishes and centrifuges.'[28]

Monsanto cemented its transformation from a chemicals company into what it now termed a 'life sciences' company by divesting itself of the majority of its older industrial assets. PCBs were long gone, as was Agent Orange. As the *New York Times* put it: 'During the 80s, Monsanto prepared for the future by sloughing off the past. From 1980 to 1987 the company sold or shut down $4 billion worth of businesses in markets deemed too cyclical: commodity chemicals and basic oil and gas exploration. In 1985 bulk petrochemicals comprised 30 percent of Monsanto's assets; by 1988 the figure was 2 percent. Last year the company sold its silicon manufacturing operation, saying it did not fit into Monsanto's plans for the future.'[29] Howard Schneiderman told the newspaper: 'There are five billion people in the world today. Some people say there should be two billion. Well, that's great. That's wishful. But that's not the way it's going to be. In the year 2030 there will be ten billion people.' As it increasingly saw itself, Monsanto wasn't just selling pesticides and seeds to farmers. It was a company on a moral mission to save the world from starvation.

In their excellent 2010 book *Fighting for the Future of Food: Activists Versus Agribusiness in the Struggle over Biotechnology*, American sociologists Rachel Schurman and William Munro report several interviews with anonymous Monsanto insiders from this period. These staffers were unanimous in stating their belief that the company's shift from chemicals to biotech was based on a real commitment to addressing environmental concerns. As one recalled: 'The scientists – well, the senior managers all the way down to the scientists – all had a very, very strong belief that we were doing something good for the world. And it was quite surprising for Monsanto that it would be seen as the opposite [of that], because everybody thought, "Here we're taking chemicals off the market and greening up the world with this new technology".'

The vision was straightforward. If the genetics of crop plants could be harnessed to directly combat pests and diseases, then agrochemicals would gradually become less important. Insect-resistant corn would not need insecticide, fungus-resistant would not need fungicide and so on. Higher-yielding crops would use less land, sparing more for nature. And nitrogen-efficient crops – or even, as a long shot, nitrogen-fixing staples like corn or wheat – would require little or no artificial fertiliser. But how to monetise shifting fewer drums of chemicals? Monsanto needed to generate an income stream from the biological intellectual property embedded in its patented GE seeds, rather than seeing crop protection chemicals as its main source of profits. 'Given its widely touted potential to help ameliorate environmental problems, biotechnology offered an apparent way for chemical companies to reinvent themselves in a new, more environmentally sensitive guise,' Schurman and Munro write.

Patents and intellectual property rights were seen as an essential component. As one anonymous industry scientist told Schurman and Munro: 'You have to have them … because it's a regulated industry and because of the heavy investment in R&D. It's the same in the pharmaceuticals and the high-tech industry ... If you're in a regulated industry and you have such a long, long lead time and huge R&D costs, you have to get a return on that. And there's not enough money in it if ... there is no intellectual property and everybody shares.' Another incentive behind taking an aggressive approach to patenting was to prevent competitors making the same discovery and then claiming ownership. 'Most of the value you get out of them is, you get freedom to operate. You get freedom to continue using something you discovered ... [Otherwise], somebody else is going to make the same discovery, patent it, and then sue you to make you stop,' said another industry scientist.[30] The enormous upfront R&D costs meant that industry scientists would only be encouraged to pursue research that could open up big new markets. Niche applications such as better crops for small farmers with limited means – especially those in poorer countries – would not get funded as they would never realistically yield a commercial dividend. This somewhat belied Monsanto's claim to be aiming to feed the world's poor.

Monsanto was well ahead of the other agribusiness majors in moving aggressively into biotechnology. 'While many other firms were active developers of the technology and important actors in the industry, no other company invested as much time, money, and human resources in establishing its position in the industry [as Monsanto],' Schurman and Munro relate. 'Nor did any other firm exert the same degree of influence over the fate of the industry and technology. Indeed, if there was one company whose name became virtually synonymous around the world with the term GMO, that firm was unquestionably Monsanto ... Unlike the other large conglomerates, which invested slowly and cautiously in biotechnology, Monsanto strove to become the industry leader from the beginning and stayed focused on that goal for the next thirty years.'

As a scientist who worked for one of Monsanto's competitor companies told Schurman and Munro: '[Monsanto executives] basically took off their gloves years ago and completely focused all of their energies in succeeding in biotechnology. DuPont, Dow, Syngenta ... all of those companies took a much more cautious approach ... Syngenta and Dow and others said, "Well, we're going to do biotech as one component of our company but still rely on the traditional part of our company to drive most of the profits." Monsanto said, "We're going to make all of our profits from biotech." And they succeeded!'

From 1975 onwards, three successive Monsanto CEOs, John Hanley (who retired in 1984), Richard Mahoney (1984–1995) and Robert Shapiro (1996–2000) 'all followed the same course, channeling an ever-increasing share of the company's resources into biotechnology and moving Monsanto away from its chemical-industry past', Schurman and Munro write. 'Shapiro was the most taken with the life science idea, betting the company's future on biotechnology in the belief that it could make the company piles of money while creating a more environmentally sustainable world.'[31]

The first fateful steps along this road had already been taken. In 1970 a Monsanto chemist named John Franz was asked to study some new compounds passed over from elsewhere in the company, which had originally been developed as water softeners. In the process of this work, Franz synthesised a new molecule called N-phosphonomethyl-glycine, better known today as glyphosate.[32] According to Monsanto's official history: 'The test results of the initial screening were so spectacular that Monsanto skipped the second screening and went straight to field trials. The first report from the trials was one word, "Eureka". The new herbicide was found to be successful in controlling both annual and perennial weeds, killing not just the leaves but the roots as well. Academic participants and growers who assisted with the field trials had one question: "Where can I buy this stuff?"'[33] Monsanto ran a 'name the compound' competition among its secretaries to come up with a brand for the new weedkiller. The winning entry came from Dottie Millis, who proposed the trade name 'Roundup'. She duly received $50 in prize money for coining a name that would become world famous and earn the company billions in revenue.

Monsanto quickly realised that it had a 'once in a century herbicide' on its hands. Roundup was very nearly the perfect weedkiller. It killed pretty much all growing plants. Its mode of action – blocking the synthesis of an essential amino acid in plant tissues – only affected the plant kingdom, meaning it had low acute toxicity in animals.* Quickly degraded by microbial action, it was more environmentally friendly than competing broad-spectrum herbicides, which often lingered in the soil or had unpleasant effects on wider ecosystems. But Roundup's very effectiveness was also its Achilles heel. Previous herbicides were useful to farmers because they were selective: atrazine, for example, was used by corn growers because it killed broadleaved weeds yet left the corn crop unscathed. However, grassy weeds resembling corn would also survive and continue to infest the crop. No conventional weedkiller could ever be 100 per cent selective in killing only weeds and leaving the crop alive.

Discovering a compound like Roundup had been a very long-term ambition for Monsanto, involving decades of research and a commitment of tens of millions of dollars. As one company insider said later: 'We started looking for what Roundup does in 1952, and we found it in '69, and we commercialized it in '75, some twenty-three years after we started looking.'[34] Such a long-term investment needed to be recouped with a successful product. Monsanto was also under increasing financial stress by the 1980s. It was rapidly divesting from traditional chemicals but had not yet built the

---

* Its LD-50, the measure of the acute dose needed to kill 50 per cent of rats in a laboratory experiment, is comparable with table salt or vinegar.

promised biotech empire to take their place. Profits were down, and senior staff began to talk of layoffs or worse. The new biotech entrepreneurs within the company, led by Howard Schneiderman and Robb Fraley, were under enormous pressure to get their new innovations to market. Their response was Roundup Ready, a dream weed control solution for farmers: all they had to do was sow seeds supplied by Monsanto that had been genetically modified to be resistant to Roundup weedkiller. With a Roundup Ready crop, the farmer could spray the growing crop with glyphosate, killing all the weeds but allowing the crop to flourish unharmed. It was to be the ultimate in herbicide selectivity. Instead of glyphosate being limited to pre-germination or marginal sprays around roadsides and field edges, it could now become the main tool for farmers to eliminate weeds, replacing pretty much all other herbicides in the process. This would help turn Monsanto's Roundup from a limited-use broad-spectrum weedkiller into a global blockbuster.

Roundup Ready had been expensive to develop. After a worldwide hunt, the necessary resistance gene had been found, appropriately enough, in *Agrobacterium* lurking in the wastewater-treatment pond outside one of Monsanto's own Roundup manufacturing plants. Monsanto had another financial motivation: it knew that Roundup was coming off patent in 2000, and that Roundup Ready could be a way to tie farmers into one seamless package, combining seeds and herbicide, expressly designed to work together for maximum weed control in the field. American farmers quickly adopted Roundup Ready after its initial release in 1996, converting the majority of corn, soy, cotton and canola commodity crops grown across the continent to herbicide-tolerant versions in the years that followed. As Schurman and Munro write, when it came to the new biotech seeds, 'U.S. farmers practically lined up to buy them. Farmers adopted these GM crop varieties faster than they had adopted any other agricultural technology in the nation's history.'[35] Perhaps the biggest benefit for farmers was that the new seeds made row crop agriculture simple. As one former Monsanto staffer wrote: 'You didn't need to be a rocket scientist to simply plant RR corn and apply 24 oz. of Roundup. Clean fields. High yields. Done. Easy peasy.'[36] Robb Fraley recalled later, 'I can remember in '96 you could drive by the soybean fields that were posted as Roundup Ready, and you could pick out the fields from the Interstate. They were clean – there were no weeds. And you'd drive along the next field and it would have lots of weeds in it. And so, it was a breakthrough and that propelled its adoption ... I think back as a farm kid walking soybean fields and pulling out the weeds by hand ... this technology has made a huge difference.'[37]

Despite its popular caricature as chemical-dependent monoculture, conversion to Roundup Ready also had environmental benefits, though it is probably a stretch to say that these were widely anticipated at the outset. It facilitated the wider adoption of no-till farming and conservation agriculture, where farmers largely stopped ploughing and drilled seeds through crop residue left on the ground. The tractors that had previously been needed for tilling and repeated weed control hoeings idled in the farmyard, leading to savings on fuel. Leaving the soil undisturbed also boosted soil carbon levels, helped reduce soil erosion and improved soil structure. Seeking to reap the publicity benefits of these carbon-saving improvements, Monsanto later adopted tackling climate change as one of its central themes, pledging in 2015 that the entire company, thanks largely to the carbon savings associated with Roundup Ready, would become carbon neutral by 2021.[38]

But although Roundup Ready was an easy sell to farmers, it was not such a hit with consumers. Farmers are Monsanto's customers, so it is perhaps understandable that the company had not thought much about how herbicide tolerance would appear to the wider public when its products began to turn up in their food. Campaign groups opposed to genetic engineering found they had an open goal – they could, with some justification, present Monsanto's seeds-and-chemicals package as a monopolistic stitch-up aimed at cementing industrial agriculture's dependence on a single chemical weedkiller. Thus Monsanto's first rollout of genetically modified crops seemed to all intents and purposes aimed at increasing, not reducing, overall pesticide sales – an especially suspicious agenda coming from the company that had brought the world Agent Orange and PCBs. Roundup Ready seemed to be a betrayal of the original aim of the biotech pioneers, which was supposedly to reduce the dependence of farmers on externally applied chemicals. This was a GMO product that by definition locked farmers onto the pesticide treadmill.

This assessment is not only evident in hindsight. It was obvious to many contemporaries that Monsanto's decision to launch genetically engineered crops with glyphosate tolerance would risk being a PR disaster. Ciba-Geigy, Monsanto's competitor (which later became Syngenta), initially decided not to go forward with its own biotech herbicide-tolerant seed traits precisely for fear of the likely popular reaction. According to Dan Charles in his 2001 book *Lords of the Harvest*, none other than Mary-Dell Chilton, Fraley's competitor in the early days of the development of plant genetic engineering, proposed herbicide-tolerant crops to her new bosses at Ciba-Geigy, only to be rebuffed. 'I can remember the immediate reaction of the Swiss bosses was, "That's an ethical problem, we'll never be able to sell that,"' Charles quotes Chilton as saying. 'They saw that it would be a problem to sell the chemical with the plants, and especially if you were trying to sell them as a package. They said, "It'll never fly; there'll be tremendous opposition to this."'

For Robb Fraley and others at Monsanto, there was no question however of forswearing herbicide resistance because of possible consumer opposition. 'I think from the first day I walked into Monsanto, the focus on Roundup resistance was a central target because glyphosate was a well-known product, it was very effective in controlling all the weeds in the farmer's field … but also killed the crop,' Fraley recalled. 'So if we could engineer a soybean or a corn or a cotton plant to be resistant to glyphosate, we knew it could be a game-changing technology for farmers.' This proved right. But although Monsanto won over farmers across the Americas, its Roundup Ready crops were entirely shut out of European agriculture because of the ensuing anti-GMO backlash. After 2000, not a single genetically engineered food crop from Monsanto or any other company was approved for domestic cultivation in the European Union*. The resulting controversy also excluded entire crop types from the long-heralded biotechnology revolution. A GM insect-resistant potato, already being cultivated in Canada, was abandoned after two warning telephone calls from McDonald's. Called 'NewLeaf', it was aimed at combating the Colorado beetle pest. When it was abandoned growers simply went back to using insecticides.[39] Herbicide-tolerant wheat, earlier considered by Monsanto to be its most promising prospect, was also shelved due to opposition from the bakers and wheat merchants. Biotech rice was dropped too.

The tsunami of opposition triggered by Roundup Ready GMOs swept away Monsanto's senior leadership and very nearly destroyed the company entirely. Its CEO, Bob Shapiro, once the apostle of biotech's bright new future, faced the ultimate humiliation, appearing abjectly at an October 1999 Greenpeace conference 'like one of those Chinese leaders who during the Cultural Revolution were made to walk through the streets in a dunce cap', as the *New Yorker*'s Michael Specter aptly put it.[40] Speaking via videolink, Shapiro 'appeared grim, defensive, and defeated', Specter

---

* A genetically engineered high-starch potato developed by BASF called Amflora was approved by the EU in 2010. This was intended for industrial applications such as papermaking, but was subject to such negative publicity that BASF discontinued it two years later.

reported. Monsanto's embattled CEO admitted that his enthusiasm for biotechnology had 'widely been seen, and understandably so, as condescension or indeed arrogance … Because we thought it was our job to persuade, too often we forgot to listen.'

It was too late for Shapiro however. In early 2000 the besieged company, its share price plunging, merged with Pharmacia (later itself bought by Pfizer) and Bob Shapiro was shown the door. The agricultural division was later spun off as a 'new' Monsanto in 2002. Today the company mostly focuses on seed sales, and – via its Seminis brand – sells high-quality vegetable seeds to both conventional and organic farmers and growers across Europe and the rest of the world. After an abortive takeover bid of Syngenta, Monsanto itself became the target of a merger in 2016 with the much larger German chemicals firm Bayer. After more than a century, the name 'Monsanto' may be about to disappear for ever.

Things might have worked out differently for Monsanto and for GMOs in general had Bob Shapiro taken the time to properly absorb the lessons of Rachel Carson's *Silent Spring*. In the book's final chapter Carson extolled the benefits of a 'bacterial insecticide' derived from the soil bacterium *Bacillus thuringiensis*. Spores from this bacterium, Carson related, contained 'peculiar crystals composed of a protein substance highly toxic to certain insects, especially to the larvae of the mothlike lepidopteras'. This *Bt* biopesticide held great promise in her view not only because it was natural and biological in origin but because of its specificity. It was non-toxic not just to mammals and fish, but also to other beneficial or non-target insects.

Carson reported encouraging early field evidence that 'the end results of [*Bt*] bacterial control were as good as those obtained with DDT' in killing caterpillars that were attacking forests, banana trees and cabbages, and without the toxic collateral damage. Applied as a spray or a dust, *Bt* was highly effective, but it did have drawbacks. It could not get to insects feeding within a plant's tissues, nor down in the roots, and it quickly degraded when exposed to open air in the field. Even so, it was effective enough to be in widespread commercial use by the 1970s and 1980s, and as a biopesticide was particularly popular among organic farmers, who were required to avoid the use of synthetic chemicals.

Genetic engineers made even better use of *Bt* by splicing the bacterial gene into the plant genome, thereby making crops express the insecticidal protein directly in their tissues. One of the earliest successes was detailed in a 1987 paper in *Nature* by none other than Marc Van Montagu, who with his colleagues detailed exactly this achievement of *Bt* expression in genetically engineered tobacco.[41] Monsanto later managed the same feat in corn and potato, and therefore might have decided in the mid-1990s to lead the rollout of crop genetic engineering with *Bt* rather than Roundup Ready. Instead the company led with Roundup Ready soy, which was first grown in 1996. The earliest *Bt* corn product, named YieldGard, was only released a year later, in 1997. So the first GMO food product launched kicked off a decades-long controversy about herbicide tolerance and whether Monsanto was simply trying to sell more chemicals, while the subsequent insecticide-reducing *Bt* seeds simply got lost in the noise. Many people got and continue to get the two things mixed up, thinking that agricultural crops had somehow been made resistant to insecticides.

My bet is that had *Bt* corn been Monsanto's initial product launch instead of Roundup Ready soy, things might have been very different for GMOs. Genetic engineering could have been associated in the public mind from the outset with the reduction of chemical pesticides and might therefore have faced less widespread opposition. Some environmental groups might even have cautiously supported GMOs as part of their long-running campaigns to reduce pesticides in agriculture. *Bt* crops might even have been adopted by organic farmers as a more efficient way to deliver a biopesticide that they had already been relying on for many years. Instead, mostly because of the 'original sin' of Roundup Ready, Monsanto found itself embroiled in a succession of controversies that have today made the company a byword for chemical-dependent 'Big Ag'.

CHAPTER FIVE

Suicide Seeds? Farmers and GMOs from Canada to Bangladesh

We all know the stories. Monsanto drives family farmers out of business. It takes away the ancient rights of farmers to save their own seeds, and it hires lawyers and private detectives to monitor and harass those it suspects of seed-saving. Worst of all, it sues farmers whose produce has been inadvertently contaminated with GMO crops through no fault of their own. You can hear all these narrative themes – of chemical pollution, farmers being sued, corporate control – in the songs on Neil Young's recent album *The Monsanto Years*. Of course as a songwriter Neil Young is entitled to a bit of artistic licence. But these claims about Monsanto are so pervasive that they are worth looking at in detail. Have farmers really lost the right to choose what to grow? Has corporate control really driven them into virtual slavery? And has Monsanto really sued farmers who have grown its patented seeds inadvertently via contamination? The company, as might be expected, flatly denies that it engages in practices like these. Monsanto states that it 'has never sued a farmer when trace amounts of our patented seeds or traits were present in the farmer's field as an accident or as a result of inadvertent means'.[1] It calls this idea a 'myth', a 'misperception … that likely began with Percy Schmeiser, who was brought to court in Canada by Monsanto for illegally saving Roundup Ready canola seed.'

Percy Schmeiser, a Saskatchewan-based farmer whose family have cultivated the same 600-hectare farm for over a century, has indeed become an international hero for those who oppose Monsanto and GMOs in general. He and his wife Louise were given the Right Livelihood Award* in 2007 for, in the words of the citation, 'their courage in defending biodiversity and farmers' rights, and challenging the environmental and moral perversity of current interpretations of patent laws'.[2] 'With their fight against Monsanto's abusive marketing practices, Percy and Louise Schmeiser have given the world a wake-up call about the dangers to farmers and biodiversity everywhere from the growing dominance and market aggression of companies engaged in the genetic engineering of crops,' reads the Right Livelihood summary.

Schmeiser was the subject of a 2009 television documentary entitled *Percy Schmeiser: David versus Monsanto*. The documentary tells Schmeiser's side of the story, claiming that 'a heavy storm at harvest-time' initially blew Monsanto's GM canola seeds onto Percy Schmeiser's fields, and despite losing the initial court case after Monsanto first sued him in August 1998, Schmeiser refused 'to be intimidated by the chemical giant' and took the case to the Canadian Supreme Court. Schmeiser, speaking about the fierce independence of Canadian farmers in standing up for their historic rights, talks passionately in the film about how he and his neighbours 'are not going to stand for a multinational to come in and try and take those rights away'. Seeing Monsanto's private investigators staking out their property, his wife Louise Schmeiser tells the camera that 'it was scary ... I felt like I was a prisoner in my own home'.[3]

The film presents the resulting Supreme Court judgement as a victory for Schmeiser, because a lower court order for large financial damages to be paid to Monsanto was not upheld. But the actual court decision tells a more complicated story. According to the *Monsanto Canada Inc v. Schmeiser*

---

* The Right Livelihood Award is sometimes referred to as an 'alternative Nobel'; 2016 laureates include Syria Civil Defence (the so-called 'White Helmets'), the Turkish independent newspaper *Cumhuriyet* and the Russian human rights activist Svetlana Gannushkina.

written Supreme Court judgement, 'Schmeiser never purchased Roundup Ready Canola nor did he obtain a licence to plant it. Yet, in 1998, tests revealed that 95 to 98 percent of his 1,000 acres of canola crop was made up of Roundup Ready plants.' Schmeiser had earlier told the Federal Court of Canada that he had first noticed the presence of Roundup Ready canola plants when he was spraying Roundup on one of his fields in order to clear ditches and spaces around power poles. Finding that a large proportion of the canola plants survived the application of weedkiller, he concluded that they must have contained the herbicide tolerance gene. However, rather than getting rid of them as other farmers had – free assistance was offered by Monsanto for this purpose – Schmeiser harvested the field and saved the seed for the following year's sowing. Unknown to him, Monsanto's private detectives, Robinson Investigations, had taken samples and sent them off to be tested following an anonymous tip-off. The tests confirmed that the canola seeds were Roundup Ready.

Although Schmeiser later claimed 'contamination', the federal judge concluded that 'none of the suggested sources [proposed by Schmeiser] could reasonably explain the concentration or extent of Roundup Ready canola of a commercial quality evident from the results of tests on Schmeiser's crop'. The judge found that Schmeiser saved seed from his 1997 crop that 'he knew or ought to have known was Roundup tolerant', and used this seed to cultivate all nine of his canola fields the following year. The Canadian Supreme Court justices consequently wrote: 'Mr. Schmeiser complained that the original plants came onto his land without his intervention. However, he did not at all explain why he sprayed Roundup to isolate the Roundup Ready plants he found on his land; why he then harvested the plants and segregated the seeds, saved them, and kept them for seed; why he next planted them; and why, through this husbandry, he ended up with 1030 acres of Roundup Ready Canola which would otherwise have cost him $15,000.' According to Monsanto, therefore: 'The truth is Percy Schmeiser is not a hero. He's simply a patent infringer who knows how to tell a good story."[4]

Despite his denials, the Canadian courts all concluded that Percy Schmeiser was trying to use Monsanto's genetic technology without paying for it. In order to justify this in court, Schmeiser challenged the validity of Monsanto's patent, claiming that living things like seeds and the resulting plants should not be the enduring property of a corporation. Both the federal and supreme courts found against him on this count too, however, stating that the Roundup Ready trait remained the property of Monsanto even in the second generation of seed. If this seems unfair, it is worth recalling that there was nothing stopping Schmeiser from continuing to grow unpatented seeds, whether conventional or organic. Monsanto had no rights over seeds in general, only those containing its patented genetic traits. Other farmers in the US and Canada who grew Monsanto's seeds, having obtained them legitimately, paid a 'technology fee' for the privilege and signed a legal document that forbade them from saving and replanting those seeds. (Monsanto has since discontinued the 'technology fee' approach and introduced what it calls 'seamless pricing'.)

Defenders of patents say they are important in order to incentivise innovation, by granting a temporary monopoly to the inventor on the commercial use of an invention. As copyright holder of this book, for example, I would seek to ensure that no one freely copies and distributes this text without my permission. Anyone doing so, as with music, software or other digital material, would be guilty of piracy. If it is true that Percy Schmeiser's Roundup Ready canola first came onto his land inadvertently, then perhaps an analogy could be made with digital music. Even if someone sends you copyrighted music by accident, you will be breaching intellectual property laws if you make further copies yourself, as Schmeiser did with his seeds once he discovered they contained the highly desirable herbicide tolerance trait. An analogy can also be made with book copyright. In his recent book about the 50 most important technologies humanity has invented, the economist Tim Harford starts by drawing attention to the copyright notice in the book's front matter. 'It tells you that while this book belongs to you, the *words* in this book belong to *me*,' he writes (his emphasis). In a similar way, even though Percy Schmeiser could own the plants and seeds that were on his farm, some of the genes inside those seeds still belonged to Monsanto.

This might seem inequitable, but patents are not permanent, generally holding for a fixed period of only 20 years. This is a much shorter duration than the typical book copyright, which in most countries runs for the author's entire life, plus a further 50 to 70 years after his or her death. The first generation of Roundup Ready is already off patent, in fact, and growers can now buy generic glyphosate-tolerant soy seeds that they can spray with similarly generic glyphosate.[5] As the company grudgingly accepts: 'Monsanto has communicated that, after the trait patent has expired, it will allow farmers to save certain Roundup Ready soybean varieties.'[6] As glyphosate too is off patent, a large proportion of the herbicide is now manufactured in China, not by Monsanto.[7] The idea is that the patent system incentivises technological innovation that, after the initial patent protection has expired, can benefit the whole of society by being widely shared.

Seeds might seem different from software or music because they are self-replicating living organisms, and therefore are not being actively 'copied' by external intervention as digital music or books might be. You might also argue that as living things they have inherent value and should not be commoditised and treated as patentable, commercially owned material in the same way as inanimate objects like iPhones or books. The ethics will always be debatable, but the legal situation – in the US at least – is clear, having been established by another Supreme Court case involving Monsanto, *Monsanto v. Bowman*. This specifically tested the issue of whether the seed's inherent self-replicability as a living organism invalidated the patent. In this case, the 75-year-old Indiana soy farmer Vernon Hugh Bowman had purchased harvested soy seeds, intended for consumption, straight from a grain elevator and replanted them, making use of Monsanto's patented herbicide tolerance trait in the process. He argued that the long-established doctrine of 'patent exhaustion' should apply in the process of second-generation seeds reproducing themselves biologically. Patent exhaustion means that you can buy and sell a patented item, a phone or a computer for example, without being in breach of the patent, which applies only in the initial instance of manufacturing and sale.

However, Bowman lost the case in a rare unanimous Supreme Court ruling handed down in May 2013. Delivering the opinion of the court, Justice Kagan wrote: 'Bowman argues that exhaustion should apply here because he is using seeds in the normal way farmers do, and thus allowing Monsanto to interfere with that use would create an impermissible exception to the exhaustion doctrine for patented seeds. But it is really Bowman who is asking for an exception to the well-settled rule that exhaustion does not extend to the right to make new copies of the patented item. If Bowman was granted that exception, patents on seeds would retain little value.'[8] To continue Tim Harford's analogy, you can sell your old books to a friend, but you can't use them as a template to make new copies of the word sequences they contain. Seeds use their genetic information as a template from which to grow a new organism. As far as US law is concerned, this biological information – assuming it is a legitimately novel innovation – can be subject to intellectual property protection just like any other invention.

This situation may be legal, but whether it is ethical or not is a different matter. As George Monbiot complained in a 1997 lecture:

> Now the genetic engineering of crop plants relies on what could be described as one-sided intellectual property rights. To ensure that they reap the benefits of their investments, the corporations that produce them are applying for, and obtaining, patents on genetically engineered crop plants. Many of the plants which the big pharmaceutical companies use as their raw material have been

developed over hundreds or even thousands of years by peasant cultivators. The corporations take them to their laboratories, play around with them for eighteen months, stick in a flounder gene here and a llama gene there, and hope to produce a lucrative new product.[9]

In other words, genetic information developed over centuries in the public realm is privatised and monopolised. This is a process that has been compared with the historical enclosure movements that dispossessed farmers in England from the sixteenth to the eighteenth centuries – except this time the enclosure is of the genetic commons rather than physical space.

Monsanto, backed by the Supreme Courts of the US and Canada, would no doubt object that it is only its inserted genes in the new genetic construct that are patented and thus privatised, not the rest of the seed genome, and that non-patented original seeds without the new trait should continue to be available. But the result, as Monbiot goes on to point out, is that patented seeds with advantageous traits tend to be more expensive: wealthier farmers are thus in a privileged position to benefit fully from them. Bigger and richer farmers may then out-compete poorer ones who do not have the means to invest in newer yield-enhancing technologies, exacerbating rural inequality. George writes: 'Small farmers, many of whom work outside the cash economy, simply cannot compete on these terms. As big producers gain access to technologies beyond the reach of the poor, they will secure an even more powerful grip of land tenure and production … It is my contention that this will result in a reduction of food security worldwide.' Thus, even though Monbiot acknowledges that higher yields should lead to more food being available – all else being equal – the political economy of patent-protected GM seeds means that in his view the opposite may happen. Whether or not this happens in the real world is disputed: one meta-analysis has shown at least that the majority of profits from GMO crops are retained in developing countries.[10] And if biotech seeds are offered patent-free and at no added cost, then the situation may be different, as we will see in the next chapter.

Many campaign groups claim that Monsanto's use of the patent system leads directly to farmers being harassed and having their rights to save seeds undermined. In a 2013 report entitled 'Seed Giants vs U.S. Farmers', the Center for Food Safety (CFS), an advocacy group that opposes genetic engineering and promotes organic agriculture, claimed that farmers 'continue to be persecuted for issues pertaining to seed patents' and that 'Monsanto has led the industry in lawsuits against farmers and other agricultural stakeholders'.[11]

Monsanto acknowledges that it does indeed prosecute farmers that it views as having breached its intellectual property rights. 'Monsanto litigation with farmers over patent infringement is a relatively rare circumstance, with 145 lawsuits filed since 1997 in the United States, or an average of 11 cases per year. To date, only nine total cases have gone completely through to a trial.'[12] Monsanto has won all the cases that came to court, and says it has donated any damages awarded to the company to its Monsanto Fund, whose mission is to provide 'sustainable assistance to communities in need around the world.'[13]

The reason that Monsanto always wins could of course be, as CFS argues, that the odds are stacked against the farmers it sues. 'Most farmers simply cannot afford legal representation against these multi-billion-dollar companies and often are forced to accept confidential out-of-court settlements,' its report claims. However, other than through the courts when a case comes to trial, there is not really any objective way to establish who is telling the truth: whether the farmer was using Monsanto's proprietary seeds inadvertently, or whether – as the courts found in the cases of both Bowman and Schmeiser – they were trying to enjoy the benefits of Roundup Ready without paying the additional technology fee.

The 'Seed Giants vs U.S. Farmers' report concludes that 'the current intellectual property regime has resulted in seed industry consolidation, rising seed prices, loss of germplasm diversity, and the strangling of scientific inquiry'. There is some evidence supporting the Center for Food Safety's case here. As with pharmaceuticals, developing a new genetically engineered crop now costs hundreds of millions of dollars, squeezing out open-source and public sector GMO innovations that could otherwise be offered to farmers patent-free for the benefit of everyone. However, these enormous costs mostly now arise because of over-strict regulation, meaning that only the deepest-pocketed corporations can afford the multi-year process of steering new crops through the byzantine approvals processes of multiple countries. Costs are also increased by delays resulting from opposition by anti-GMO groups like the Center for Food Safety, which generally opposes, through court action or lobbying regulators, every new genetically engineered product, irrespective of whether it is private or public sector-derived. Indeed by raising costly barriers to market entry for open-source, small business or public sector innovations, these anti-GMO groups have ironically helped to cement the very same corporate consolidation of the seed industry that they cite as a justification for their opposition.

There is actually one example of Monsanto fighting a battle with US organic farmers, but in this case it was the organic farmers who sued Monsanto, not the other way around. This began in 2011 when OSGATA, the Organic Seed Growers and Trade Association, filed suit in a New York district court. OSGATA's court deposition began by stating that 'Coexistence between transgenic seed and organic seed is impossible because transgenic seed contaminates and eventually overcomes organic seed.' Consequently, it feared that growers 'could quite perversely be accused of patent infringement by the company responsible for the transgenic seed that contaminates them'.[14] Therefore, 'Plaintiffs ask the Court to declare that, should they ever be contaminated by Monsanto's transgenic seed, they need not fear being sued for patent infringement.' However, both the district and appeals courts ruled that OSGATA did not have a case because Monsanto had already undertaken not to sue anyone for inadvertent contamination. To my mind the whole thing smacked more of a public relations exercise than a serious legal challenge. The US Supreme Court later declined to hear the case, allowing the lower court rulings to stand.

One of the biggest concerns regarding corporate control in the GMO debate is the widely held notion that Monsanto's genetically engineered seeds do not reproduce – that they are intentionally sterile, and that this forces farmers to return year after year to purchase seeds from the same company. Dubbed 'Terminator technology', this is often given as a reason why genetic engineering is necessarily bad for farmers, and why private corporations seem so determined to push it. There is indeed something intuitively offensive about the idea of sterile seeds, that biological reproduction itself could be genetically switched off through human technological manipulation, all for corporate profit.

It is just as well then, perhaps, that seed sterility as a trait – though it was proposed and partially developed in the 1990s – was never actually deployed anywhere in the world. So the much-vaunted 'Terminator technology' does not actually exist. The story that Monsanto's seeds don't reproduce is really a myth. Like most myths, however, there is a kernel of truth at the heart of it: Monsanto did purchase a seed company called Delta & Pine Land that in the 1990s had been involved in developing non-reproducing seeds. Ironically, one of the motivators behind the original development of so-called 'Genetic Use Restriction Technology' was to eliminate the danger of inadvertent genetic contamination, as well as the more commercial drive to protect intellectual property. Those worried about Terminator also tend to forget that hybrid seeds, which have been around for nearly a century, don't breed true in the second generation either and so must be bought anew by farmers each year. However, in response to global outrage Monsanto later pledged not to use the technology. Today, therefore, all GM seeds (unless they are hybrids) reproduce just like any other – which is precisely why Monsanto has chased nearly 150 farmers through the US courts to stop them being replanted without the company's consent.

A more justifiable criticism of Monsanto is that it has achieved a dominant position in the transgenic seed market, thereby raising anti-trust and monopoly concerns. In the decade after the launch of its Roundup Ready system in 1996, Monsanto went on a spectacular spending spree, snapping up almost 40 companies. Most of these were either biotechnology or seed companies, including Agracetus, Calgene, Holdens, Asgrow, DeKalb Genetics and Delta & Pine Land as well as Cargill's seed businesses. The other four of the 'Big Five' agrochemical companies – DuPont, Syngenta, Bayer and Dow – did likewise. Between them, these companies own virtually the entire market in transgenic seeds, and Monsanto traits are in more than half of all the seeds sold. The picture is complicated because from the outset Monsanto took a 'broad licensing' approach, allowing competitor companies to use its patented biotech traits in their own seeds. However, the US Department of Justice was recently sufficiently concerned as to open an investigation into 'possible anticompetitive practices in the seed industry'. However, this investigation was quietly closed in 2012.[15] Why? 'In making its decision, the Antitrust Division took into account marketplace developments that occurred during the pendency of the investigation,' was all a Department of Justice spokesperson would tell Tom Philpott, a reporter for the magazine Mother Jones.[16]

Despite the US Department of Justice's refusal to get involved, corporate concentration in the seed sector has not gone away; in fact, it is getting worse. Dow and DuPont have merged into DowDuPont and as mentioned earlier Monsanto is itself being acquired by Bayer. Syngenta has been bought by ChemChina, and the merger of Monsanto with Bayer will reduce the Big Five to the Big Three. 'The acquisition activity is no longer just about seeds and pesticides but about global control of agricultural inputs and world food security,' warned the technology watchdog ETC Group (Action Group on Erosion, Technology and Concentration), in a 2016 press release.[17] In July 2017 a coalition of the American Antitrust Institute (AAI), Food & Water Watch and the National Farmers Union wrote to the Department of Justice specifically asking for the Monsanto-Bayer merger to be blocked on competition and innovation grounds.[18] AAI president Diana Moss said: 'The merger substantially eliminates competition across a number of important markets. It could squeeze out smaller rivals and saddle farmers and consumers with higher prices, reduced choice, and less innovation.' The letter points out that if the Monsanto-Bayer merger goes ahead, the resulting combined firm will have $26.9 billion (almost £20 billion) in revenue, 40 per cent of combined industry revenue, and will be larger even than DowDuPont and Syngenta-ChemChina.[19]

Whether competition regulators will allow the largest of these mega-mergers to be completed remains to be seen. If they do, protecting genuine competition in the transgenic seeds sector – surely important both to encourage further innovation and to defend the rights of farmers against high prices – will become an increasingly serious concern. This is not the same as aiming to ban or heavily restrict GM crops; instead it is more about seeking to ensure that the technology is not unduly restricted by corporate control. The groups campaigning against corporate dominance on GMOs are somewhat disingenuous here in my view. Food & Water Watch and ETC Group are not trying to open up competition and increase the ability of farmers to access novel innovations like genetically improved seeds: they are trying to restrict or even ban farmers from using GMOs at all. Their professed concerns about anticompetitive practices therefore seem more tactical than genuine to me. Still, corporate concentration is a real issue and the recent spate of mega-mergers will probably make things worse both by restricting choice for farmers and by giving more ammunition to anti-GMO activists intent on demonising big corporations for maximum propaganda value against the technology.

Of all the accusations levelled against Monsanto, the claim that the company is responsible for the deaths of hundreds of thousands of Indian farmers is surely the most serious. The allegation about Indian farmer suicides has been repeated by countless newspapers, given worldwide exposure by award-winning documentary films and even mentioned in speeches by Prince Charles. It gives a strong moral drive to anti-GMO campaigners who believe they are defending the rights of some of the world's poorest and most vulnerable farmers. And it cements the reputation of Monsanto as being one of the most ruthlessly exploitative and villainous corporations in the world.

'The GM genocide: Thousands of Indian farmers are committing suicide after using genetically modified crops,' read the headline of one heart-rending story published in the British tabloid the *Daily Mail* in 2008.[20] The article contained first-hand accounts of farmers who had killed themselves by drinking insecticide, leaving grieving families to pick up the costs of debts and failed crops. 'Here in the suicide belt of India, the cost of the genetically modified future is murderously high,' the reporter concluded after visiting Maharashtra state. Another first-hand story of a farmer suffering crop failure was the centrepiece of the 2011 film *Bitter Seeds*. 'Every 30 minutes a farmer in India kills himself,' declared the poster produced for the movie. The documentary showed in over a hundred film festivals, received the 'Global Justice Award' from the development funder Oxfam Novib, and was aired on dozens of TV channels internationally. The *New York Times* writer Michael Pollan called it 'a tragedy for our times, beautifully told, deeply disturbing'. The website for the film described the problem: 'The GM seeds are much more expensive; they need additional fertilisers and insecticides and must be re-purchased every season.'[21]

Perhaps the most vocal campaigner against Monsanto is the Indian environmental activist Vandana Shiva. In a long 2014 profile by the *New Yorker* writer Michael Specter, Shiva was described as 'a hero to anti-GMO activists everywhere' for her 'fiery opposition to globalisation and to the use of genetically modified crops'.[22] The US television correspondent Bill Moyers has described her as a 'rock star in the worldwide battle against genetically modified seeds'. Shiva wrote about the Indian farmer suicide issue on her website in 2016 under the heading 'Monsanto vs Indian Farmers'. In this article she claimed that cotton seed had been 'snatched from the hands of Indian farmers' by Monsanto, and that the entry of the latter into the Indian seed market had increased the price of cotton seeds by '80,000%' (no, that is not a misprint). Most importantly, she alleged that as a result '300,000 Indian farmers have committed suicide, trapped in vicious cycles of debt and crop failures, 84% of these suicides are attributed directly to Monsanto's *Bt* cotton'.[23] Elsewhere, Shiva has called this a 'genocide', one she insists was directly caused by Monsanto.[24]

There is much more to this widely circulated story, however, than first meets the eye. *Bt* cotton contains an insect resistance gene which means that it inherently requires less, not more, pesticide. Instead of having to spray the crop with insecticides, farmers should be able to rely on the cotton plants being resistant to the main insect pest, the cotton bollworm. So why should farmers be getting into debt through a need to purchase 'additional pesticides', as the *Bitter Seeds* movie alleges? Another obvious question is why farmers would act against their own apparent interests by purchasing the same 'failed' GM cotton seeds year after year. *Bitter Seeds* puts this down to aggressive marketing on the part of the 'seed salesmen' acting on behalf of Monsanto. Likewise, the *Daily Mail* writer reported that 'GM salesmen and government officials had promised farmers that these were "magic seeds"'. But *Bt* cotton was first widely planted in 2002. Now, 15 years later, it still accounts for over 90 per cent of Indian cotton acreage, with 800 different competing *Bt* cotton varieties on the domestic market. Have farmers really been conned by clever marketing into buying useless 'magic seeds' 15 years in a row? Are Indian farmers really such perennial victims that they fall for the same tactics year after year and end up having no recourse but to kill themselves? As the saying goes: fool me once, shame on you; fool me twice, shame on me. But fool me 15 times? That would make Indian cotton growers surely the dumbest people in the world, the lemmings of world farming.

The real-world evidence suggests, not surprisingly, that Indian cotton farmers are not stupid at all, and that they freely choose to grow *Bt* cotton because it increases their yields, reduces costs from insecticide and brings them and their families more money. The most rigorous fieldwork examining this question was conducted by Jonas Kathage and Matin Qaim of the University of Goettingen, Germany, and published in the prestigious journal *PNAS* in 2012.[25] Kathage and Qaim surveyed 533 farm households over four cotton-growing states (Maharashtra, Karnataka, Andhra Pradesh and Tamil Nadu) between 2002 and 2008. Among the farmers they surveyed, *Bt* cotton adoption rose from 38 per cent in 2002 to a spectacular 99 per cent in 2008, suggesting that either Monsanto's 'magic seed' salesmen were incredibly persuasive or, more likely, the farmers decided that *Bt* cotton delivered real benefits. This is exactly what the German researchers found, with a 24 per cent increase in cotton yield due to reduced insect pest damage and a 50 per cent gain in profit for *Bt* adopters, largely because of these higher yields. In a separate paper, Qaim and his colleague Vijesh Krishna showed that *Bt* cotton had led to a 50 per cent drop in the use of pesticides over the same area and time period – a significant benefit both to the environment and to the health of farmers.[26] Qaim estimated that if the benefits of reduced pesticides are extrapolated to India as a whole, '*Bt* cotton now helps to avoid at least 2.4 million cases of pesticide poisoning every year.'[27] That should be news of the sort to warm any environmentalist's heart – even that of Vandana Shiva.

And – although they never seem to get interviewed by well-meaning visiting documentary film makers – there are plenty of Indian farmers prepared to tell this more positive story. One I met who left a lasting impression on me was Gurjeet Singh Mann, a cotton farmer from the western state of Haryana, resplendent in his red turban. I was impressed at his quiet humility and strong commitment to the environment when I met him in Delhi. 'Prior to *Bt* we tried every kind of lethal poison that was available on the market to spray on our cotton crops,' he said later when speaking to a colleague of mine from Cornell University.[28] 'Every evening we were spraying our fields with insecticides. This charged the environment with poisonous fumes that created havoc with the bird life, animal life, insects, frogs, sparrows, and they were in no time vanished from our villages. You could no longer hear chirping of birds.' It is a story that sounds very reminiscent of Rachel Carson's *Silent Spring*. As Singh Mann reported, since the widespread adoption of *Bt* cotton and the resulting decrease in pesticide sprays, 'we have again chirping birds near our villages, our national bird peacock has arrived back, pigeons are there, insects we can see again, we can see frogs during the rains, so animal kingdom was also returning to normal after the advent of *Bt*'.

So why, if *Bt* cotton has not been the calamitous failure that is so often reported, are Indian farmers apparently killing themselves in such large numbers? The suicide issue has been exhaustively researched by Ian Plewis, a professor of social statistics at the University of Manchester in the UK. Examining official suicide rates, Plewis found that non-farmers were more likely to commit suicide than farmers in six out of the nine Indian states that extensively grow cotton. 'If anything,' he writes, '[annual] farmer suicide rates, at about 29 per 100,000 are a little lower than for non-farmers (35 per 100,000)' across the whole cotton-growing region.[29] In other words, India's 'suicide belt' exists as much in its cities as on its farms. The absolute number of suicides looks big to outsiders because there are simply so many farmers in India; 40 million in the nine cotton states alone. What matters therefore for comparative purposes is not absolute numbers of suicides but the rates per unit of population. Reassuringly, these show that suicide rates in India are not much different to those in other countries. The annual Indian farmer suicide rate is higher than in England and Wales, Plewis reports, but 'similar to the best estimates of the rates in Scotland and France'[30] (neither of which, I might add, currently grow GMO crops).

Plewis also compares suicide rates before and after the introduction of GM cotton. If *Bt* cotton were driving a 'GM genocide', one would expect a leap in suicides after its widespread adoption. But this is not what the data shows. 'In 2001 (before *Bt* cotton was introduced) the suicide rate was 31.7 per 100,000 and in 2011 the corresponding estimate was 29.3' – actually a small reduction.[31] Plewis concludes, in contrast to the widely believed Indian farmer suicide story, that 'the pattern of changes in suicide rates over the last 15 years is consistent with a beneficial effect of *Bt* cotton, albeit not in every cotton-growing state.'[32] So not only is the ubiquitous *Bt* cotton-suicides story incorrect, but 'there is evidence to support the hypothesis that the reverse is true: male farmer suicide rates have actually declined after 2005 having been increasing before then'. The Indian farmer suicide story is a myth built on tragic individual anecdotes and extrapolated to a whole country by those like Vandana Shiva with an ideological axe to grind and little concern about the true facts.

Perhaps the more interesting question then for *Bt* cotton adoption in India is how the outside world got the reality so wrong. As I suggest above, I don't think this happened by accident. To use modern terminology, the *Bt* cotton suicide myth seems to be on closer examination a classic piece of 'fake news'. It is especially curious to ponder why environmental campaigners have been so implacable and vociferous in opposing an innovation that has demonstrably reduced pesticides and pesticide poisoning of farmers. As it happens, I gained some first-hand experience of how easily false narratives can be generated on GMOs through later work I did in Bangladesh, with Cornell University, the government-run Bangladesh Agricultural Research Institute (BARI) and the United States Agency for International Development (USAID).

With USAID funding, Cornell and BARI had collaborated with the Indian seed company Mahyco to take Monsanto's *Bt* genes (donated by the company free of charge) and insert them into local South Asian varieties of aubergine (eggplant, or brinjal, also called *begun* in the Bangla language). This *Bt* brinjal was originally intended for three countries: India, Bangladesh and the Philippines. As this was a mainly public sector and philanthropic endeavour, farmers – unlike their better-off compatriots in North America, or indeed Indian farmers growing *Bt* cotton – would be given the seeds royalty-free. These seeds would not be separately patented but – via government-owned research institutes – would remain the property of the farmers themselves to save and share with friends and neighbours as had always been the case. Rather than there being new varieties of *Bt* brinjal, the genes were introduced into existing farmer-preferred varieties of aubergine, seven in total, with local names like Uttara, Kajala and Nayantara. This allowed farmers the same choice of local varieties as always, just with new genetic pest protection.

As with *Bt* cotton the overall aim of the project was to address the rampant overuse of pesticides. Brinjal is an important vegetable in South Asia, but is attacked by a caterpillar pest called the fruit and shoot borer. In order to protect the crop from decimation by these voracious moth larvae, farmers have historically been forced to spray as many as 80 to 140 times during the growing season. Human exposure to toxins can be very high as a result: farmers usually spray bare-footed and typically wear no hand, eye or face protection. Chemical insecticides used in the region, which include various organophosphates and carbamates, tend to have higher toxicity than those employed by Western farmers owing to a laxer regulatory and enforcement regime. As a result of widespread exposure, more than a quarter of farmers surveyed in one study reported experiencing multiple health effects, including headaches, eye and skin irritation, vomiting or dizziness attributable to pesticide use.[33] Long-term health effects associated with these insecticides have been found to include non-Hodgkin's lymphoma, leukaemia, birth defects and cancer.

In a rational world, environmental groups would therefore have been keen partners in the promotion of a pesticide-reducing crop. I visited numerous Bangladeshi farmers, sometimes in remote areas of the country many hours' drive from the capital Dhaka, and found that all of them had dramatically cut their use of insecticides, sometimes right down to zero. Yields were higher, the vegetables looked better without insect damage, and were popular when sold by farmers in local bazaars, often proudly identified with hand-written labels as 'insecticide free'. Yet when I visited many farmers, I found that anti-GMO activists had got there first, and had already tried to convince Bt brinjal growers that their new crop was poisonous merely because it was genetically engineered. One particularly pernicious rumour spread by these activists was that Bt brinjal would make farmers' children paralysed if they ate it. Instead of growing Bt brinjal, they advised farmers to either switch to organic – in which case most of their crop would probably be destroyed by pests – or to go back to spraying toxic insecticides.

This experience taught me how quickly a 'failure' narrative could catch on if it served the purposes of determined and highly ideologically motivated campaign groups. At the time I started working on the Bt brinjal project, stories were already appearing in the Bangladeshi press claiming that the crop had failed, that the new Bt plants were dying in the fields, and that angry farmers were demanding compensation and vowing never to plant these dreadful GMOs again.[34] 'The cultivation of genetically engineered Bt Brinjal in the country's several districts has cost the farmers their fortunes again this year as the plants have either died out prematurely or fruited very insignificantly compared to the locally available varieties,' read the first sentence of one such story

published in March 2015.[35] This wasn't completely false: some
of the farmers I visited had indeed seen a failure of their *Bt*
brinjal crops. On closer investigation though, this turned out
to be because of an outbreak of bacterial wilt, showing
unsurprisingly that *Bt* brinjal was as prone to bad weather,
bad luck or poor cultivation practices as any other crop. Some
of the negative press stories claimed that the *Bt* trait had failed
and that fruit and shoot borer insects had attacked the
genetically resistant crop. When I visited I also found pest
damage – but only in the control crop, which was typically
planted around the sides of the genetically engineered plants
for comparative purposes and as a long-term strategy to help
forestall the evolution of insect resistance. The most likely
explanation therefore was that activists had simply
misidentified what they were looking at and had confused the
GM and non-GM brinjal plants. I suspect they already knew
what they wanted to find and so didn't take the trouble to
check the facts.

In some cases individual farmers were quoted by anti-GMO
campaigners as bemoaning lost *Bt* brinjal crops, only to tell
the exact opposite story when we checked back in with them.
One such farmer was Mohammad Hafizur Rahman, whom I
met in 2015 on his farm in Tangail district, north of Dhaka.
As he handed around delicious slices of watermelon, we sat in
his two-roomed house and discussed his farm. He proudly
showed us around his fields, as local children gathered and
chattered around us. I mentioned his positive experience,
with nearly doubled yields, and much less pesticide use, in a
later article I wrote for the *New York Times*.[36] This must have
made him a target, even though I purposefully didn't reveal
the exact location, because an anti-GMO journalist and
accompanying activists later visited him and, according to
Rahman, handed over literature against *Bt* brinjal. The
farmer later told a Bangladeshi colleague of mine, Arif
Hossain from the Bangladesh Alliance for Science, 'They
gave me a book and told me that, "Look brother, *Bt* brinjal
has various problems". They also told me not to eat this
brinjal. They said if insects don't eat this, it must not be a

good thing for humans to eat. With my practical mind to counter them, I asked them that people take medicines for worms, the worms die, why don't people die? They were not able to answer my question.'[37]

The activists and journalists also wrote that Rahman's crop was dying, and claimed it as an example of a failure of the GMO *Bt* brinjal crop. But the farmer was aghast at this when my colleague Hossain brought it to his attention. Instead, he insisted that the activists had simply not understood that the crop had already been repeatedly harvested and had reached the end of its growing season. 'When he [the reporter] visited me, those plants had started to die. The plants had no brinjals on them, I had already started to harvest Dhundul [Sponge Gourd] in that field. So, I told him my harvesting of brinjal is finished already.' Had he told the anti-GMO activists that he was dissatisfied with *Bt* brinjal? 'I did not say this. When the plant comes to the end of its life, it dies. My *Bt* brinjal plants died when they were finished fruiting. Everything comes to an end, doesn't it? Will the brinjal plants stay throughout the year? That's not possible.' To understand how the *Bt* cotton failure story became so well-established, imagine testimonies such as that of Mohammad Hafizur Rahman's, repeated many thousands of times to credulous journalists or activists determined to find evidence to support a strong pre-existing narrative.

Working from spring 2014 for Cornell University as one of the partners in the *Bt* brinjal project, I found myself fighting pitched battles with anti-GMO campaigners across multiple fronts. There are numerous blog posts, videos, news reports and tweets from this time if you want to follow the sequence of events.[38, 39, 40] In retrospect it might have been better for the farmers themselves to have been able to tell their stories directly to an international audience rather than to have their views mediated by outsiders. But at least the Bangladeshi aubergine cultivators were given the opportunity denied to so many farmers elsewhere around the world: to decide for themselves whether to grow genetically modified crops. The fact that they have so far made a success of them shows that they are, like farmers everywhere, just as capable as anyone else of taking their own decisions about what is best for them. If the *Bt* brinjal seeds do indeed fail at some point in the future – as the activists claim they must – then no doubt the farmers will just as quickly give them up. The crucial thing is that this should be their choice, not mine or Greenpeace's.

The stakes were particularly high for the scientists and activists in Bangladesh because everyone knew that *Bt* brinjal was the world's first genetically modified food crop produced in the public sector for the use of small farmers in developing countries. Thus it was a world away from Monsanto's herbicide-tolerant monocultures in the US prairies. Those convinced that genetic engineering was an intrinsically bad technology were therefore determined to see it fail with *Bt* brinjal in Bangladesh, while those promoting the use of this technology were equally keen to have a success story to point to. The issue was especially politicised because activists in India had succeeded in forcing the government to declare a moratorium on *Bt* brinjal in 2010, stopping the project from moving forward there, and in fact holding up all biotech approvals since that date. Similarly in the Philippines, Greenpeace and others had both destroyed *Bt* brinjal test sites and obtained legal injunctions against its use (although a Philippines Supreme Court judgement against *Bt* brinjal was reversed in July 2016).[41] Successful deployment in Bangladesh would have upended those groups' campaigns by showing that farmers in a developing country can make use of genetic engineering after all.

Let's be clear about the real-world impact of this activism however. Aubergine farmers in both India and the Philippines have sprayed millions of pounds of additional insecticides thanks to the activities of Greenpeace, Vandana Shiva and other anti-GMO campaigners and groups in denying them the opportunity to grow *Bt* brinjal. This will have harmed the ecology of farmers' fields in both countries, as well as the surrounding environment and water resources. And it will have resulted in thousands or even tens of thousands of unnecessary cases of pesticide poisoning in farmers and agricultural labourers. This is what happens when ideology trumps science – the environment is harmed, people get sick, and some even die.

In October 2016 activists and campaigners from around the world gathered in The Hague, Netherlands, for an unusual event. The location was purposefully selected: The Hague is the venue for the International Criminal Court, which tries cases of crimes against humanity such as those against the perpetrators of alleged mass human rights abuses and genocides in Darfur and Libya. However, these campaigners were not converging on the real International Criminal Court; they were setting up their own tribunal to judge Monsanto. The International Monsanto Tribunal would, they hoped, serve as a venue to air grievances about the company's actions in a number of areas. In its own words, the tribunal aimed: 'To examine the effects of Monsanto Company's activities on the human rights of citizens and on the environment, and to offer conclusions about the conformity of Monsanto's conduct with the principles and rules of international human rights law and humanitarian law.'[42]

Although I didn't attend the event, I did watch its proceedings with interest. In writing this book, I was eager to find some recent solid evidence of wrongdoing by Monsanto, because I was worried that the material I had found so far seemed to present too much of a positive judgement on the company. Surely Monsanto was not 'the most evil company in the world' only because of historical crimes committed back in the 1960s and 1970s and GMOs? Even though the tribunal did not seem very balanced to me – it was organised by IFOAM (the International Federation of Organic Agriculture Movements), Navdanya (an NGO established in India by the anti-GMO campaigner Vandana Shiva), the Organic Consumers Association (a US NGO that also promotes anti-vaccination) and others[43] – at least some of the evidence given might point me towards real-world recent examples of harm.

Although the judges seemed to be bona fide human rights lawyers, the International Monsanto Tribunal did not behave like a normal court. It did not hear evidence from both sides of the argument because there were no witnesses for the defence. Monsanto itself refused to have anything to do with it, writing an open letter stating its opinion that the tribunal was a 'staged event, a mock trial where anti-agriculture technology and anti-Monsanto critics play organisers, judge and jury, and where the outcome is pre-determined'.[44]

And on this, Monsanto had a point. The tribunal, in the preamble to its findings, made the following rather odd statement:

> The Tribunal has no reason to doubt the sincerity or veracity of those who volunteered to testify before it. But, because their testimony was not given under oath or tested by cross-examination, and because Monsanto declined to participate in the proceedings, the Tribunal is not in a position to make findings of fact concerning the allegations of various company misdeeds. Rather, for the purpose of answering the questions posed for the Tribunal's consideration, the Tribunal will assume that the facts and circumstances described by the witnesses would be proven.[45]

The case for the prosecution was thereby bolstered by the fact that no witnesses appeared for the defence and all prosecution witnesses were assumed always to be speaking the truth, thereby proving their own cases by definition. In addition, the verdict would not itself be based on 'findings of fact' about Monsanto's alleged misdeeds. This was a strange legal procedure if ever I heard one. But still, what about the specifics of the testimonies heard by the tribunal? One of the witnesses was Percy Schmeiser, whose case I described in detail above.[46] Another was an Australian organic farmer called Steve Marsh, who had sued his neighbour Michael Baxter – a conventional farmer growing GM canola – claiming contamination and the loss of his organic status.

The case was a long-running cause célèbre in Australia, but like Schmeiser, Marsh lost in the Supreme Court of Western Australia.[47] The judge wrote that the extent of the 'contamination' was just eight volunteer canola plants ('volunteer' meaning plants not sown by the farmer). Marsh did not grow canola, so cross-pollination was not an issue, and he could just have pulled the self-sown plants up and thrown them away.[48] Marsh lost the case because according to the court what he was trying in effect to do was to extend the special sensitivities of his organic status over his neighbour's property, which he had no right to do.

Another of the witnesses was Farida Akhter, leader of the campaign against *Bt* brinjal in Bangladesh, who I had already tangled with personally in that campaign. As well as alleging that *Bt* brinjal had failed in the majority of farmers' fields – a claim I knew from direct experience to be false – she also presented unlikely claims about *Bt* brinjal's supposed health effects, including cancer, infertility, liver damage, allergies, 'unpredictable mutations' and other such scientifically unsupported nonsense.[49] Reading through this testimony I was shocked by the idea that all of this would simply be 'assumed to be true' in what purported to be a legitimate judicial process. It was also odd to have this featured in the Monsanto Tribunal because Monsanto is not involved with the *Bt* brinjal project, other than to have donated its *Bt* genes free of charge at the outset, so any alleged failures of *Bt* brinjal were nothing to do with the company.

Another witness was Gilles-Eric Séralini, whose infamous paper, claiming to show that rats fed GMO corn and Roundup developed tumours, was retracted by the journal *Food and Chemical Toxicology* after it was roundly panned by virtually the entire scientific community for questionable statistics. Most of the other witnesses – a pig farmer from Denmark and witnesses from Colombia and Paraguay – were testifying about what they claimed were health impacts from glyphosate. This is another area of enormous controversy; the pig farmer presented graphic images of piglets born with hideous deformities, but his claim that these were associated with

GMO feed crops and glyphosate residues lacked any scientifically credible justification. Not passing muster in a highly regarded scientific journal, the allegations had been relegated to a fringe pay-for-play journal published by a company that has been widely identified as 'predatory'.[50]

Many of the witnesses cited health problems they claimed were caused by glyphosate. These were supported by references to a judgement by the International Agency for Research on Cancer (IARC) of the World Health Organisation (WHO), issued in 2014, that glyphosate is 'probably carcinogenic to humans.' That sounds damning, but IARC is a peculiar body – its review processes are questionable, and its judgement on glyphosate was opposed by all other scientific agencies in the world, including the European Food Safety Authority, which concluded that 'glyphosate is unlikely to be genotoxic (i.e. damaging to DNA) or to pose a carcinogenic threat to humans.'[51] After this, the two agencies embarked on a vicious war of words.

The IARC's judgement on glyphosate was also controversial because a former employee of the US group Environmental Defense Fund called Christopher Portier gave evidence as an 'invited specialist' to IARC's committee on the issue.[52] It was revealed in October 2017 that Portier had failed to declare on several occasions that after the IARC process had concluded he had been paid $160,000 by a US-based legal firm hoping to profit from a class-action lawsuit it would then undertake on behalf of the 'victims' of glyphosate poisoning, suggesting a conflict of interest.[53] The news agency Reuters has also revealed that several conclusions on which IARC based its assessment of carcinogenicity had been suspiciously altered between drafts of its report. 'Reuters found 10 significant changes that were made between the draft chapter on animal studies and the published version of IARC's glyphosate assessment. In each case, a negative conclusion about glyphosate leading to tumors was either deleted or replaced with a neutral or positive one,' Reuters revealed. Its article was damningly titled 'In glyphosate review, WHO cancer agency edited out "non-carcinogenic" findings.'[54]

Either way, IARC's judgement on cancer risk overall needs to be put in perspective. While glyphosate is placed in category 2A of 'probably carcinogenic', it shares this designation with red meat, wood smoke, manufacturing glass processes, drinking 'very hot beverages over 65°C' and even the occupation of being a hairdresser. In the higher category 1 ('carcinogenic to humans') you will find numerous unsurprising villains such as plutonium and tobacco smoke, but also sunshine, soot, salted fish ('Chinese-style') and processed meats such as bacon.[55] Even if the IARC's judgement is taken at face value, it hardly supports the allegations by Monsanto Tribunal witnesses that glyphosate has caused birth defects, kidney failures and numerous other ailments, for which no credible scientific evidence has ever been presented.

Elsewhere in the tribunal, I had more sympathy with a witness from Colombia, who gave moving testimony about the spraying of glyphosate on coca crops by the US and Colombian governments as part of the 'Plan Colombia' drug eradication programme. In my view, there is little doubt that aerial herbicide sprays were an abuse of human rights, killing legitimate crops as well as coca, and driving small farmers into poverty and off their land. Thankfully the programme was discontinued in 2015,[56] but I would lay the blame with the governments responsible for the spraying. In addition, there is no guarantee that Monsanto was even the manufacturer: these days most glyphosate is made generically by Chinese companies. I guess the Zhejiang Wynca Tribunal or Sichuan Fuhua Tribunal (both Chinese companies manufacturing generic glyphosate) doesn't have the same ring to it.[57]

There were also testimonies about Monsanto's lobbying activities and how these might influence governments. Clare Robinson from the UK-based group GM Watch presented a witness statement about how Monsanto had been involved in lobbying to establish a GMO regulatory system that in her view was insufficiently rigorous.[58] She also cited some WikiLeaks cables showing US officials strong-arming GMOs in battles with the EU, but these were not specifically about Monsanto, nor did they show much evidence that Monsanto was directly involved.[59] Robinson further referenced a 2005 case where Monsanto was fined for bribing an official in Indonesia, which was covered in the media at the time.[60] This all seemed pretty sketchy to say the least.

I realise I am perilously close to doing the opposite of the tribunal here and appearing to exonerate Monsanto by demolishing the tribunal's witnesses. All I can say is that the evidence presented to their tribunal by some of the key anti-Monsanto activists in the world was remarkably thin, being based largely on anecdotes, contested allegations and some-times outright pseudoscience. I was left thinking: is this really the best they can do, these people who spend their lives battling this corporate behemoth? Wouldn't it have made a stronger case to ensure the testimony was factually accurate rather than just making the procedural assumption at the outset that all Monsanto critics would be telling the truth? Of course in the end the tribunal found Monsanto guilty, as it was always going to do. But any multinational in the world would have been judged harshly in such a process. Would Google have fared any better? Or Apple? Or even Amazon-owned Whole Foods, which has a similar annual turnover to Monsanto, of about $15 billion (£11 billion)? All large companies, by dint of their size alone, wield power that can outstrip that of elected governments, and without sufficient scrutiny and accountability this power will end up being abused. Monsanto, however, isn't even in the top 50 big corporations,[61] appearing at number 189 in the 2016 Fortune 500 list,[62] having slipped a long way down from its ranking of 33 back in the good old chemical industry days of 1965.[63] Maybe Agent Orange was better business than GMOs after all.

Perhaps the best – or worst – that can be said is that Monsanto is a key part of an agricultural system that a lot of people find objectionable, and sometimes for good reason. But GMOs of any sort, and the companies that develop and promote them, need to be looked at in the context of the political system that they are deployed within. While Roundup Ready soy might have helped privilege bigger farmers in North and South America, this experience would not necessarily be repeated with *Bt* brinjal in Bangladesh, nor with other genetically modified crops being developed specifically to address poverty in Africa and elsewhere. Yet too few anti-GMO campaigners seem able to recognise the crucial difference of context, conflating all genetically modified crops with Monsanto, monoculture and pesticides, as did the witnesses at the Monsanto Tribunal.

Personally I would agree with Oxfam America that the most rational approach is to look at each issue case by case. Oxfam sensibly states that is has 'no policy position for or against GM technology',[64] and continues:

> *Oxfam believes that any decision to use GMOs must be based on the human rights principles of participation, transparency, choice, sustainability and fairness. Feeding the world's hungry people requires huge social, political, economic and cultural changes, not a simple technological fix. Oxfam understands that technology does matter and that modern biotechnology might play a role in helping to achieve global food security, but only so long as farmers are central to the process and that their rights are strengthened, not harmed.*

This sounds good to me. However, as the next chapter demonstrates, farmers in developing nations in Africa have mostly not been able to access modern biotechnology, even when doing so would most likely strengthen both their rights and their food security. And the culprits have not been big multinationals like Monsanto, but NGOs who are supposed to be promoting, not harming, the interests of the world's poor.

CHAPTER SIX

Africa: Let Them Eat Organic Baby Corn

It was early August 2013, just a few months after my Oxford anti-GMO apology speech. I was bumping along a pot-holed coastal road in a dust-streaked vehicle with Dr Joseph Ndunguru, a plant virologist and one of Tanzania's most respected scientists. Dr Ndunguru had earlier shown me the labs and greenhouses he worked in, on the outskirts of Dar es Salaam, and now he wanted to demonstrate the real-world context of his efforts. He told me that Tanzania was at the epicentre of a devastating viral outbreak – one not affecting humans directly, but which instead was destroying a source of staple food for rural Tanzanians, a crop called cassava.

Also known as tapioca or manioc, cassava has a deep bunch of long tubers with a bushy shrub growing above ground, and seven-fingered green leaves sprouting from woody stems. As a source of carbohydrate it is one of the most important crops throughout sub-Saharan Africa. It is drought-tolerant and hardy, growing pretty much anywhere on poor soils and with minimal inputs of fertiliser and water. In times of drought, the roots can be left in the ground, or dug up and dried for future use. Cassava is a true subsistence crop, one that will reliably get a family through times of hardship when other sources of food disappear. If the cassava too failed, as it was now beginning to do, many subsistence-level Tanzanians would have nothing to fall back on. This is what Dr Ndunguru wanted to show me.

Not far from the coastal town of Bagamoyo, we pulled off the road and up a track, past mud-walled thatched huts scattered under sparse eucalyptus trees. I got out of the car and looked around in the baking heat. Under a rudimentary shelter of coconut palm leaves, a woman in a patterned headscarf was cooking something in a blackened pot balanced on three rocks over a smoky fire. Another woman with a baby was sitting on a rattan mat. She wore a yellow Unicef T-shirt, and as I approached, the baby cried out in fear and hid away in her bosom. Tanzania is a country where a third of the child population is stunted by chronic malnutrition. Lack of healthy food is the greatest contributor to its tragically high rate of infant mortality, with 130 child deaths every day resulting from extreme poverty. The country has 2.7 million stunted children by the most recent count.[1] Stunting is not just a passing phase of childhood; by impairing brain development it can permanently affect a child's life chances, perpetuating poverty.

Dr Ndunguru called me over. He was examining a group of rather miserable-looking cassava bushes. As he turned over their leaves, I saw that they were yellowing and shrivelled, indicating the disease-stricken state of the plants. Ndunguru explained that the symptoms were for both cassava mosaic and cassava brown streak viruses, which tended to strike together in a deadly partnership. He pulled up one of the plants and split the root apart with a penknife. It was streaked with brown rotten areas. Clearly there would be little to harvest from this crop. A glum-looking farmer approached us, wearing a dirty red vest and ragged brown trousers. Ndunguru asked him to demonstrate how much food his family had left. He went into a nearby hut and reappeared with a blue plastic plate on which he had placed a few dried-up pieces of scrappy cassava. That was all.

The farmer was Ramadhan Issa, aged 35, though to me he looked much older. He had three children, he told me in Swahili with one of our scientist colleagues interpreting. The eldest was nine, the second was six and the youngest just two. They were all down to one small meal a day, just cassava, and sometimes maize flour from the market. I asked about the situation of the cassava crop on his meagre 1.5-acre land-holding. 'We don't expect to get much,' he replied. He wasn't sure whether it was drought or disease, but either way the crop was failing. If he uprooted the whole farm, he told me, he might expect to harvest one single bag of cassava. That would be about 100 kilogrammes to last the family for months. In a normal harvest they might expect five bags, or half a tonne. So what would they do? 'We will wait for the rains and replant,' he replied. That was all they could do.

Issa told me what his life was like during these difficult times. 'You are hungry and you can't work, so you can't make any money to buy food from the market.' The whole family was now suffering malnutrition-related health problems, he said, especially the children and his elderly parents, who were also living with the family and similarly dependent on what Issa could grow and harvest. Clearly there was no chance of any of the family getting a balanced diet in this state. His focus was merely on getting through each day. They were down to a daily ration of two small pieces of cassava each, he told me, equivalent to about half a potato.

We walked over to the next homestead, where a woman in a white singlet stood with her three children. She told me her name was Grace Rehema, a 25-year-old mother of three children. I lined them all up to take a photo, the children in front of the shrunken, virus-infected cassava plants they were supposed to depend on. How much food did they have? Nothing much – the remains of a sack of maize flour. No cassava even, she told me. Where would the next meal come from after that? She didn't know.

Dr Ndunguru's project was breeding genetically modified cassava with genes designed to make it resistant to these devastating viruses. In neighbouring Uganda I had already seen some successful trials of virus-resistant cassava; there the plants looked lush and healthy, especially in comparison with the shrivelled and infected cassava being grown in this part of Tanzania. However, as things stood, even if Dr Ndunguru's cassava proved successful in the laboratory there was little chance that farmers like Ramadhan Issa and Grace Rehema would be allowed to grow it. Strongly influenced by European attitudes, Tanzania had strict laws against GMOs that prohibited scientists like Dr Ndunguru from even being able to test their disease-resistant crops in the fields, let alone share them with needy farmers.

The following day involved a gruelling eight hours on the road, heading west towards Dodoma, Tanzania's government centre. At each fly-blown stopping place, where minibuses loaded with locals would pull over to pick up passengers, children offering groundnuts and bananas would tout their wares. Women squatted by the side of the road selling melons or vegetables, but in this arid part of the country there was nothing like the diversity of food I had seen in Uganda and other more lush parts of Africa. Poverty was endemic: after darkness fell there was no sign of electric light from the homesteads I knew lined the road and the surrounding areas. Occasionally our headlights picked out children, often dressed in Maasai tribal colours, shading their eyes from the glare of passing traffic while herding goats and sheep by the roadside.

Outside Dodoma – which despite being the administrative capital of Tanzania retains something of a dusty frontier feel – was an agricultural research station where government plant scientists had optimistically prepared the ground for a 'confined field trial' of genetically modified crops. I visited it the next morning, shaking hands with the director and conducting other formalities before the obligatory group photo on the steps outside under the relentlessly hot African sun. In a spare few minutes I sat down at a wooden desk in one of the meeting rooms with Dr Joseph Ndunguru and two other leading scientists, Dr Alois Kullaya and Dr Nicholas Nyange, to try to get a better understanding of the situation in the country.

Dr Kullaya told me that he was the country coordinator for WEMA, the Water Efficient Maize for Africa project supported by the Bill & Melinda Gates Foundation, aiming to make drought-tolerant maize available to small farmers in several countries in East Africa.* Although Kenya, South Africa, Uganda and Mozambique were all moving ahead, he told me, the 'strict liability' provision against GMOs in Tanzanian domestic law was blocking the next step for WEMA, which would be to field test drought-tolerant genes in local maize varieties at the proposed outdoor field site nearby. 'Strict liability' originated in a model law for Africa promoted by the United Nations Environment Programme under the Cartagena Protocol on Biosafety, a process strongly influenced at the negotiations stage by anti-biotech campaign groups and delegates from Europe. Under normal concepts of liability, Dr Kullaya explained, you might be liable for damages if you were negligent or failed to take precautions against some foreseeable harm. With strict liability, any claimed damages could be applied to any party in the entire chain of development for a genetically engineered crop on pretty much any pretext. 'Whether or not I took all the precautions that I could as a scientist, or a technology developer, I will be found guilty,' Kullaya lamented. Moreover, 'that covers all those who are in the chain, whether the technology developer, the person who transported the seed, the one who financed it, even the director who approved the project will be part of it.' And the penalties were daunting. All those named could find themselves paying substantial damages to any anti-GMO group claiming harm, or could even be sent to prison. This was not meaningful regulation, Kullaya insisted, it was de facto prohibition. The law had been designed by outsiders not to facilitate the use of the technology in a safe and responsible way, but to stop it being used in Africa at all.

---

* WEMA was being developed as a project by AATF, the African Agricultural Technology Foundation, which aims to get better crops, developed using modern methods, into the hands of farmers without them paying royalties. AATF supported my visit to Africa in 2013, when I visited Kenya, Uganda, Tanzania, Zimbabwe, Ghana and Nigeria.

Next to speak up was Dr Ndunguru. His disease-resistant cassava was already performing well in the lab, he told me. So how long would it be until it could be available to benefit farmers like Ramadhan Issa and Grace Rehema? Just a few years, Ndunguru replied. But he was facing the same problem as the maize scientists trying to develop drought-tolerant corn. 'The strict liability thing is also limiting our research,' he complained, 'because we need to move that technology out of the lab and take it to confined field trial and eventually to commercialisation.' That could never happen with such a prohibitionary legal regime, which treated healthy virus-free crops as a 'biosafety' threat in the same way as you might a germ warfare programme. Unused to fighting political battles, the scientists were baffled and frustrated.

In the meantime, I was told, the anti-GMO groups were dominating the national conversation. Just the previous day a newspaper had carried an article in Swahili by an opponent of genetic engineering, warning farmers not to plant GMOs because of the dire consequences he said would necessarily follow. Such articles were typically full of misinformation, complained Dr Nicholas Nyange, a scientist working for the Tanzanian government's Commission for Science and Technology and responsible for promoting science in the country. Money was flooding into Tanzania from rich countries, but it was not going to support farmers in accessing improved crops. It was going to groups who promoted only organic farming and agro-ecological 'alternative' agriculture of the type now fashionable in Europe and among many donor agencies. The recipient NGOs, which he named as PELUM (Participatory Ecological Land-Use Management) and the Tanzanian Organic Agriculture Movement, were 'working with smallholder farmers, poor farmers, in encouraging them to use saved seed, farmer-saved seed, traditional seed'.

Well, what could be wrong with that? I asked. Surely everyone agrees that farmers saving seed is a good idea? Yes, Dr Nyange countered, 'but how can you support it if it is poor-yielding seed? How can you support a traditional seed while you know that traditional seeds are inferior in productivity?' According to him, low-yielding traditional seeds left farmers trapped in a cycle of subsistence, where they barely produced enough harvest to get through the year. 'The outcome is insufficient food supply, food insecurity in the country.' Instead, he said, 'we want the farmers to change so that they can grow for themselves and also have enough that they can sell and make an income out of that. We want the farming activity to be much more profitable to our farmers … so that they can have enough for food but they can sell and have income so they can take their children to school, they can afford better health services, they can have all that we want.'

But why? Maybe farmers preferred traditional lifestyles, and the NGOs were right to defend them? 'There is no farmer who wants to remain poor!' Nyange responded passionately. Just basic hybrid maize seed, not even the drought-tolerant GMOs in development as part of WEMA, could bring a step-change improvement in productivity, he insisted. 'We want them to have that alternative to use improved seed, to have more yield and to increase the productivity. There is nothing wrong with a farmer if instead of harvesting five bags of maize from traditional seed, they are harvesting twenty bags of maize from improved seed.' And the traditional landrace varieties would not be lost, Nyange reassured me, as they were all being collected and preserved in the government's crop genetic resource centre, not least as a future source of genetic diversity for plant breeders to draw on.

Dr Nyange paused. All he was asking, he insisted, was that some of the poorest farmers in the world should not be condemned to suffer the lowest harvests because they were denied the choice of growing crops using better high-yielding seeds. At the very least, farmers should surely be given the choice to decide for themselves. It did strike me as somewhat peculiar that NGOs, which were ostensibly concerned with poverty reduction, should doggedly oppose such a basic poverty-reducing measure as better seeds for farmers, even non-GMO ones. These activist groups, it seemed, were intent not just on opposing the use of genetic engineering in Africa but also on blocking the introduction of modern farming methods in general, including such basic necessities as fertiliser, weed control techniques and improved seeds. And the losses were mounting, Dr Ndunguru pointed out. Droughts were getting worse thanks to climate change. Insect pests were ravaging crops and spreading diseases. There were more whitefly – the main insect vector for cassava brown streak and mosaic viruses – than ever before. 'For brown streak you are talking of 35 to 70 million US dollars annual losses,' Ndunguru said. 'For cassava mosaic you are talking of four million metric tonnes annually from the country. It's a huge loss. And to us such crops are the major food security staples.'

But maybe conventional breeding could help, as the opposition groups insisted it could without the use of genetic engineering? 'We have been trying to address these issues using conventional methods without success. So we would like to use this [GMO] technology.'

'And to those who say that Africa is not ready to use this technology?' I asked.

'They are not Africans,' Ndunguru replied. 'What is the alternative? If you ask these people who say we are not ready to use it, "what is the alternative solution to these situations where you get extremely poor yield", they have no alternative.'

'There's no agro-ecological solution to cassava brown streak?' I persisted.

'No. It's not there. With banana bacterial wilt, it's not there. With maize lethal necrosis [another emerging disease], it's not there. Cotton diseases, it's not there.'

I was struck by the complete divergence of world views between the scientists I was speaking to and the campaigners I had spent many years of my life with and still counted as friends. The former viewed crop diseases and low productivity as problems to be fixed with better seeds and tools, just as they had been elsewhere around the world. The activists were not interested in such pragmatic solutions, which were likely to be dismissed as 'techno-fixes'. They had a different vision of agriculture, one based on a rejection of modern farming methods such as hybrid seeds, mechanisation and the use of fertilisers and pesticides. However, this deference to traditionalism seemed questionable in Africa, where it might mean children labouring in the fields, and women and girls walking miles to gather firewood and fetch water rather than attending school. Although the scientists were keen to point out to me that they respected the agro-ecological principles advanced by the NGOs, they were clearly sceptical that those promoting such techniques really understood the challenges of reducing poverty. 'They should not really deceive the farmers that they can produce profitably and make a better life by using ecological farming,' Dr Kullaya insisted. 'And you know I think it is these people who do not know what it means going hungry to bed.'

'Are any of these NGO people subsistence farmers themselves?' I asked.

'I don't know even if they are farmers. I think these are people who are being employed. Whoever they are being paid by, I doubt whether they even do farming themselves. So to deceive farmers that you can live from ecological farming agriculture, I think that is not proper.'

'Not proper? I would use a stronger term than not proper,' I suggested. 'I would say it is morally reprehensible.'

'Well,' Dr Kullaya responded thoughtfully. 'English is not my mother tongue but you have put it correctly. These guys do not know what it means to go hungry to bed. There are people, farmers where we are operating in Dodoma, where they harvest almost nothing because of the drought, because of the diseases. We know biotechnology is not a silver bullet, but it is part of the solution to what the farmers are experiencing.'

As it happened, I had already got a direct taste of just how morally reprehensible the extremes of anti-GMO activism could be on my journey from Dar es Salaam to Dodoma the previous day. We had stopped at a halfway point, in Morogoro, at a roadside hotel on the edge of a small national park called Mikumi. The area is hilly, with a forest-cloaked mountain standing as a prominent landmark to the south of the small town. The town was also the venue for an important agricultural show, meaning that there were lots of organic activists in town showcasing their ideas. Some of them, wearing identikit yellow T-shirts with an NGO logo printed in black on the front, attended a public talk I was invited to give in a conference room at the hotel.

As I outlined my own evolving views on the GMO issue and answered questions, one of the yellow-shirted organic activists stood up and announced something in Swahili. Everyone froze, and an embarrassed titter spread through the room. It wasn't until the meeting broke up that I was able to ask one of the scientists, Dr Ndunguru, what the activist had said. 'What he said was that some genes have been implanted in this [GMO] maize so that when our kids eat it, they will be abnormal. Instead of behaving as a male, they will be behaving as a female. This will lead to homosexuality.' He gave an embarrassed snort.

'So it was kind of a homophobic comment,' I said.

'Yes.'

'I don't get it,' I replied. 'Where do these ideas originate?'

Ndunguru shrugged. 'People say that. You know, this false information spreads like fire.'

Another of the scientists spoke up, adding: 'They said that with this maize, the next generation will have some sexual deformities, their children would have homosexual tendencies as a result of eating this maize. It's really weird.' They both seemed at a loss for how to respond. This clearly wasn't the sort of discussion they were used to having as scientists.

In neighbouring Uganda things were if anything even worse. I found that some big-name charities had been spreading myths and conspiracy theories about GMOs that had poisoned the political debate and paralysed the legislative process for years. Since 2012 parliamentarians from all major parties had been trying to pass biosafety legislation to ensure that genetically modified crops, once developed, could be distributed to farmers with proper testing and safeguards. Anti-GMO campaigners from internationally recognised charities and development agencies had fought a pitched battle not to amend or improve the proposed National Biotechnology and Biosafety Bill, but to prevent it from being passed at all. This, they reasoned – correctly, it turned out – would lead to a situation of permanent paralysis which would be just as effective in preventing the use of genetic engineering in Ugandan agriculture as any formal prohibition.

I heard at first hand how fractious this debate had been from one of the senior parliamentarians tasked with taking the proposed bill forward in his role as vice-chair of the parliamentary Committee on Science and Technology. In a side room in Uganda's parliament building in downtown Kampala, ruling party MP Robert Kafeero Ssekitoleko related how activists from supposedly reputable organisations had been targeting his constituents directly in order to prevent him voting for the Biosafety Bill. 'Now they are creating so many unnecessary fears,' he told me, 'like for example they can say, if you want a long banana, you can pick a gene from a snake and put it into a banana so that the banana becomes the length of the snake.'

I couldn't help but laugh. People really took such nonsense seriously? Absolutely, Kafeero told me – in fact, on one occasion, anti-GMO activists had used religious sentiment to stir people up almost to the point of violence. 'One scientist, backed of course by some activists, went to my constituency. They mobilised Muslims, and told them "look, they are going to get a gene from a pig, to put it into maize, to make that maize fat, as fat as a pig!" [For Muslims a pig gene would presumably be unclean.] And all my constituents became rowdy to me, they say, "you are propagating this, now look what you are bringing to us!"'

'And you actually felt physically threatened?' I asked.

'Of course. It was serious on my side because it would make me lose an election in the future. So what I did, I picked up a team of scientists who also went to counter that. So we did it scientifically. And it looks like the farmers are understanding now.'

Kafeero also turned the tables by inviting the original proponent of the pig genes myth to repeat the allegation under oath in parliament. 'When he came to my committee he swore [the oath]. And after swearing, he denied everything. He changed the story completely. So if he can't say it in our committee under oath, then why does he say it elsewhere?'

Undaunted, the anti-biotech activists changed tack, planting stories in the media accusing Kafeero and his fellow MPs on the parliamentary Committee on Science and Technology of taking $30 million (£22 million) in bribes from Monsanto. They also seeded new conspiracy theories that GMOs 'have come to destroy our people, to reduce our life expectancy'. Unfortunately, even some cabinet ministers found this new idea persuasive. 'I was arguing with one person yesterday in cabinet,' Kafeero told me. 'I was like, OK, just think about it scientifically. If somebody says GMOs are here to reduce our life expectancy, USA is the leading producer of GMOs, and is actually the leading consumer of GMOs, but their life expectancy is over 70 years. Ours in Uganda is a miserable 47 years. So if those feeding on GMOs last longer, then why are you saying they are coming to kill us here?'

Anti-GMO agitation had affected MPs from all parties in Uganda, I discovered. One opposition MP I spoke to was Beatrice Atim Anywar, the member of parliament for Kitgum district in the northern part of the country. She represented Uganda's main opposition party, the Forum for Democratic Change. Beatrice Anywar was universally known as Mama Mabira, after the Mabira forest that she fought to save from being converted to sugar plantations in 2007. She won her campaign, and although still shrinking today the Mabira forest is still home to endangered primates such as the Uganda mangabey, as well as leopards and numerous rainforest bird species. Her campaign involved bringing thousands of protesters onto the streets of Kampala, earning the ire of Uganda's authoritarian president Yoweri Museveni in the process. After three people were killed in the melee,[2] Mama Mabira was thrown into jail accused of terrorism. Even today, she told me, being an opposition politician is always an 'uphill task' in Uganda where 'you have to seek the permission of the police in order to go and consult with your own constituents'.

Despite having risked her life to save one of Uganda's forests from destruction, Mama Mabira has now come under attack by environmental activists for saying that she supports the use of genetic engineering in agriculture. 'I believe that biotechnology in the situation of food insecurity is the answer because we need to increase our yields, and we need to come out of poverty,' she told me. 'We need to use biotechnology on our land to get the best out of it.' Because Uganda has high population pressure and only a small land area, she insisted, increasing yields was also imperative to protect remaining rainforest areas like the Mabira forest from future conversion to agriculture.

Anywar's experience of anti-GMO activism was similar to that of her parliamentary colleague Kafeero. She told me about her experience of attending one meeting in the northern town of Gulu, where photoshopped images were displayed to farmers. 'They make false photos, which to the ordinary Ugandan who is not computer literate, can imagine that it's true, they go around saying that when you eat GMOs you are going to produce children with the heads of corn plant,' she complained. I later obtained a copy of the offending image Mama Mabira mentioned. It was a strange montage showing sick-looking babies' heads emerging from ears of corn as flocks of black crows flapped ominously overhead in a dark sky.

When I asked Kafeero for names, he mentioned ActionAid, as well as the Catholic aid agency Caritas and an NGO called the Food Rights Alliance, and claimed they were all participating in anti-GMO activism. I told both Kafeero and Anywar that ActionAid is a reputable charity in the UK, and that I had even donated to it myself, as had others I knew, because we supported its mission to reduce poverty. 'What they are doing right now, the money they are having, they are using it to campaign on GMOs and to frustrate the farmers not to undertake the biotechnology,' Anywar responded. She was worried that Europe was perhaps aiming to keep Africa food insecure to maintain old colonial relationships of dependence, 'which to me as an African is unacceptable … and my call is that whoever is funding ActionAid or related international NGOs must stop as we also hold them responsible for this campaign.'

I also found clear evidence of ActionAid's complicity in spreading misinformation in Uganda. This came in the form of a recording of a radio advert that was being broadcast nationally across the country. 'This is a message from ActionAid!', the Ugandan-accented voice announced confidently. 'Did you know that GMOs can cause cancer and infertility?' I promised MPs Kafeero and Anywar that I would do what I could to expose ActionAid's behaviour back in its home country, the United Kingdom. Subsequently, when the GMOs-cancer advert was revealed by both the BBC and the *Independent* newspaper a year later, an embarrassed ActionAid head office in London quickly disavowed the advert, apologised and undertook not to spread similar anti-scientific myths in future. So far, I am pleased to say, they have kept to this commitment.

This might count as a win for science, but ActionAid is only one of dozens of groups to have been actively spreading anti-GMO fears across numerous African countries. These groups also frequently claim to speak on behalf of African farmers. One petition circulated in 2013 by the African Biodiversity Network (an anti-GMO group concerned less with actual biodiversity than with banning biotechnology) purported to speak on behalf of '400 organisations across the African continent representing farmers, indigenous peoples and civil society groups' in a call to 'Ban GMOs in Africa'.[3] But when I spoke directly to actual legitimate farmer representatives in Kampala in late 2013 they expressed anger and frustration about externally funded NGOs speaking on their behalf. Willie Bamutiire, elected representative for the Uganda National Farmers Federation representing the Kampala district, told me that the NGOs are easy to spot.

'You can see them driving in posh cars, and calling conferences,' he complained, referring not specifically to the African Biodiversity Network but to NGOs in general. 'These are conferences that are never attended by farmers, or farmer representatives.' What Bamutiire found especially galling was to switch on the radio and hear news reports that 'farmers have rejected [GMOs], yet farmers have never been consulted'. These NGO people 'just masquerade as farmer representatives', he told me, but they can be very effective. 'They are sponsored. They have the money so they can go anywhere, they can reach anywhere.'

The MP Kafeero had already told me about the grassroots modus operandi of these NGOs. 'What they do actually is they organise seminars and workshops and bring farmers, farmer groups, they facilitate them transport and everything, they give them food. They keep them there the whole day and inform them the negative about the GMOs.' In town after town, district after district, these events would include the usual parade of anti-GMO speakers warning of cancer, infertility and pig genes, while also making presentations using misleading photoshopped images.

What farmers were after, the Uganda National Farmers Federation representative told me, was pretty much the same as I heard in Tanzania, Kenya and other African countries. 'What our farmers want is hybrid seeds ... We need to increase our productivity,' both to reduce poverty and to adapt to climate change. They certainly didn't want outsiders telling them that they didn't need to use modern farming methods.

Ugandan farmers confirmed that their crops, as in Tanzania, were suffering from outbreaks of new diseases. Particularly threatened were bananas, which are picked green and eaten mashed and steamed as a staple food called *matoke*. About a third of the bananas grown globally are produced in East Africa, where they provide a quarter of the daily food requirements for a population of more than 100 million people in Uganda, Burundi, Rwanda, Tanzania and Kenya. Uganda is the world's second-largest producer of bananas thanks to *matoke*; the vast majority of these bananas are produced on a small scale for household and local consumption.

In 2001 a previously little-known bacterial disease spread across the border from Ethiopia into Uganda and began infecting banana plantations. Spread by flies and infected tools, Banana Xanthomonas Wilt (BXW) manifests in a rapid wilting of the plant and a rotting of unripened fruit bunches. Cut open an infected stem and a yellow pus-like substance oozes out, while the whole tree eventually turns black and dies. Farmers whose plantations were affected had no choice but to slash down and bury all infected trees and those nearby. As in Tanzania with cassava, households depending on banana as the main staple of their family meals were threatened with renewed food insecurity because of the BXW epidemic. As with cassava, international scientists working in partnership with government-run African institutions were racing to develop resistant banana varieties using genetic engineering.

While in Kampala I met one of the lead scientists on this effort, Leena Tripathi from the International Institute of Tropical Agriculture. Banana is very much an indigenous crop, Tripathi told me. 'It is grown at a household level, so the farmer grows in the garden,' she added, as we shared a lunch of *matoke* and beans together at Makerere University in Kampala. 'There are only very few in commercial production, and very little goes into the export. It is mainly for their consumption and the small market.' This was why a threat to banana was such a direct concern in terms of malnutrition and food insecurity in Uganda and its neighbours. Even after an exhaustive search, plant scientists had not been able to find any resistance genes in banana or its close relatives, so had instead used transgenic technology to import a gene from sweet pepper.

First results were promising, Tripathi told me. She had found 100 per cent resistance in 11 new lines of genetically transformed varieties. We later looked around her greenhouse, where she showed me adjacent potted banana plants: one was the variety with the new resistance gene, and the second a control without the gene. The resistant variety was robust and vigorous. The control was dead. So how long would it be, I asked Tripathi, before farmers would be able to obtain the resistant variety of banana? She hesitated. 'Before that, at the moment, Uganda doesn't have the biosafety law, so the law needs to be in place before these varieties can be commercialised,' she said slowly.

I was left with the impression that doing the laboratory work was the easy bit. Without the long-delayed biosafety and biotechnology law, Tripathi's wilt-resistant bananas would stay locked in the laboratory indefinitely, even as the disease made steady progress across the Ugandan countryside and millions of households faced renewed hunger as a result.

While in Uganda I did discover one pro-science NGO among the many groups dedicated to stopping crop biotechnology. Conveniently, SCIFODE (Science Foundation for Livelihoods and Development) also shared a building with the Ugandan National Farmers Federation in Kampala. SCIFODE's communications director Peter Wamboga took me out to visit the National Crops Resources Research Institute (NaCRRI), a government research station in the small town of Namulonge. It was here that I saw the virus-resistant cassava varieties being grown in field trials. As we travelled north on bumpy roads, our driver expertly navigating around potholes, goats and people, Wamboga twisted round in the front seat and regaled me with his view of the injustice of the current situation.

'There's no doubt that the money is coming from Europe,' he complained. 'Why can't Europeans send money to Africa to buy tractors? Why don't they send money to buy hoes, to buy tools that can reduce on the drudgery of harvesting and delivering of crop yield to homes? Why should Europeans send money to fight advancement of Africa in terms of technology?'

'The NGOs say they want to support traditional smallholder agriculture,' I suggested. 'What's wrong with that?'

'Let them support traditional agriculture in Europe,' Wamboga countered. 'Is that what they are doing? Are they growing traditional crops in Europe? They are growing improved crops. They tell Africa, grow these old small traditional crops. Europe is contributing largely to African under-development.'

'But what about those who say that Africa isn't ready for these new technologies?' I asked (this was a refrain I had heard many times from activists).

'It's absolute nonsense. What do they mean by not ready? Europe does not want Africa to be food secure in order to be independent. Europe was Africa's coloniser. It does not want Africa to be fully independent. They still want to control our ability to have food or not to have food.'

I told him that I doubted that even the most fervent anti-biotech NGOs saw their role as neo-colonial. In fact, they claimed to be combating the colonialism they saw as inherent in the importation of improved crops and the resulting penetration of multinational seed companies into traditional agriculture.

'Who is saying it? Is it Africa saying, or is Africa parroting what Europeans say? We are not sub-human,' he went on, his voice rising with indignation. 'We are fully human beings, with functional human abilities and capacities, and we can decide for ourselves.'

As a European, I was beginning to feel uncomfortable. But Wamboga was not finished. 'Science is a universal knowledge and resource. Nobody can claim to own, understand, know better what science and technology is. Who told us to adopt mobile phones? Mobile phone penetration is highest in Africa on the globe. Nobody told us. And Europeans are not telling us to take up mobile phones or don't take mobile phones. Why should they tell us in agriculture – "take this technology, don't take this technology". Why should they choose for us, why should they decide for us? Why should they mislead Africa?'

The conversation was broken by our arrival at NaCRRI, where a red-shirted young Ugandan researcher took us out to a field some distance from the low-rise buildings that housed the institute's laboratories and meeting rooms. We paused in front of an intimidating-looking gate with spikes on the top, on which was pinned a large sign announcing in black and red capital lettering: 'FIELD TESTING SITE FOR GENETICALLY ENGINEERED PLANTS'. Through this, down a slight slope, was another sturdy wire fence and a second gate, this one held firmly shut with a chain and a strong-looking padlock. Another large sign announced, this time in green capital letters: 'G.M. CASSAVA PLANTS. FOR RESEARCH PURPOSE ONLY'. And in smaller letters underneath: 'Not approved for food or feed. AUTHORISED PERSONNEL ONLY.'

The NaCRRI researcher opened the padlock and with some trepidation we all entered the forbidden field. The view was impressive. Laid out in front of me was an acre of the healthiest-looking cassava plants I had ever seen. Beaming proudly as I took photos with my phone, the red-shirted researcher plucked two leaves and held them out side by side. One was yellowing and shrivelled, indicating signs of viral infestation. It was a non-GM control plant, he told us. The second leaf was dark green and robust – a virus-resistant transgenic trial plant. There was no sign of mosaic or brown streak virus on this variety, he told us. All around, the healthy bushes swayed in a warm wind, anchored firmly by the swelling tubers on their roots into Uganda's deep-red soil.

I asked the scientists at NaCRRI how long it would be until these resistant cassava varieties might be released to farmers. There were still some technical problems in handling the complex relationship between the two different viruses, I was told, but the main challenge had nothing to do with science. Instead, I heard once again that the major logjam was getting parliament to pass a biosafety law, so that genetically modified crops could be distributed within a proper regulatory framework. This gave me an idea. What if, I asked our hosts quietly, someone were to spirit a bit of resistant cassava root out of the padlocked field, hand it over to a local farmer and start the ball rolling that way? The answer was unambiguous – a large fine or ten years in jail. Not worth the risk.

These images stuck in my mind for a long time afterwards. The padlock seemed to be something of a metaphor for the state of African crop science, a permanent symbolic barrier between the smallholders waiting in their millions for better crops, and the plant scientists and crop breeders who were tasked with developing them. Could anyone unpick this political padlock? The process has been desperately slow given the urgency of the situation, but there are optimistic signs. Thanks largely to the work of Wamboga and other pro-science advocates within Uganda, the national parliament finally passed the Biosafety Law on 4 October 2017. I had just left the country having spent a week in Kampala and visiting banana farmers across the equator in Mbarara district. I knew that they would all be celebrating. On one farm I had watched the farmer cut down and burn infected banana trees. I thought of him when I read the news about the success in parliament.[*]

Everywhere I went in Africa it was the same story. Foreign-funded NGOs, supported mainly by donors in Europe, were delaying or blocking the development not just of biotechnology but of modern agriculture generally across the continent. I visited Kenya in 2013 and again in subsequent years, to find the situation deteriorating each time I went to Nairobi. In

[*] Although, just as this book was being finalised, in late December 2017, the Ugandan president – perhaps under the influence of anti-GMO campaigners – refused to sign the Biosafety Bill and sent it back to Parliament. Two steps forward, three steps back, it seems.

2013 a newly minted biosafety agency was up and running, and scientists had initially expected insect-resistant maize and disease-resistant sweet potato to be approved for cultivation soon. I had a chat with the director and some of his staff, and was surprised to find them twiddling their thumbs as they waited for the first applications to be submitted.

The reason for the delay could as usual be traced back to Europe. In 2012, the French professor Gilles-Eric Séralini (whom we have already met as a witness at the Monsanto Tribunal) published a paper purporting to show that GMO-fed rats developed tumours. Few other experts took his work, which was published together with colour photographs of the hideously disfigured animals, seriously, but in Kenya, activists saw an opportunity. They managed to convince the then health minister Beth Mugo, who had recently suffered from breast cancer, that GMOs might have been the cause of her disease. In tears, and brandishing the now infamous tumour-riddled-rat photos at a subsequent cabinet meeting, Mugo persuaded Kenya's president to announce an immediate ban on GMO imports. The ban was promulgated without much semblance of due legal process, with no reference to Kenya's new biosafety authority and against the objections of the nation's scientific community; instead it was announced in a triumphant press conference given by Mugo herself.[4] Even though Séralini's offending paper was later retracted,[5] Kenya's ban remains in force today. Without its removal, there is no realistic prospect of approvals for any of the improved biotech crops that have been developed by national scientists to help the country's farmers.

Each African country has its own unique experience of anti-GMO fearmongering. In Ghana, which I visited late in 2013, field trials were then underway in the northern district of Tamale for a *Bt* variety of cowpea (better known internationally as black-eyed pea). The lead scientist showed me round, explaining how cowpea was the most important protein crop in the whole of West Africa, and was especially crucial as many people in rural areas could rarely afford meat or other animal protein. He also told me how farmers were forced to spray pesticides or risk losing their crop due to infestations of an insect pest called the pod borer. The *Bt* cowpea would be resistant, so farmers were keen to start growing it. This was a similar story to *Bt* cotton, corn and aubergine. The first trials had proven promising, with higher yields and no sign of the offending insect.

As in Bangladesh, however, even the prospect of a public-sector genetically engineered crop getting into the hands of smallholder farmers was opposed by anti-GMO groups. In 2015 an NGO called Food Sovereignty Ghana (FSG) applied to the Ghanaian high court for an injunction against both *Bt* cowpea and a separate nitrogen-efficient, drought-tolerant rice also in development in the country.[6] The case was launched with a noisy protest behind a large 'March Against Monsanto' banner, even though Monsanto would not be selling the *Bt* cowpea seeds – the work had been done in the public sector by a consortium of African and international scientific and philanthropic organisations.[*] FSG's demand for prohibition was in keeping with its long-standing policy for, as it put it, a 'total ban on everything GMO, including but not limited to, the introduction into the environment, contained and confined use or field trials, import, export, GMO in transit, or placement on the market'.[7]

According to its website, Food Sovereignty Ghana's opposition to GMOs is because of 'increases in birth defects, severe depression, precocious puberty in girls, autism, childhood cancers, male sperm quality and infertility, Parkinson's disease, cardiovascular disease, diabetes and chronic kidney disease'. Needless to say, no genuine scientific evidence supports any of these assertions. Although FSG's legal action eventually failed, the court case dragged on for months and threw a cloud over the *Bt* cowpea project, generating

---

[*] *Bt* technology is owned by Monsanto, so the company is listed as an official partner in the project in some communications. For the purposes of *Bt* cowpea, it has been licensed to the African Agricultural Technology Foundation, based in Nairobi, for royalty-free disbursement to farmers. The work is supported by the Bill & Melinda Gates Foundation.

numerous press articles where activists were given free rein to hype the health dangers supposedly inherent in GMOs.

In Zimbabwe I found the situation even worse, with anti-GMO hysteria elevated to national policy by then-President Robert Mugabe's dictatorial government. One of Mugabe's cronies was Tobaiwa Mudede, who had penned a series of bizarre articles about the supposed dangers of GMO foods. In one diatribe published by the Zimbabwean *Sunday Mail*, Mudede claimed: 'Further research has established a link between GMO consumption and induction of such diseases as inflammatory bowel disease (IBD), colitis, autism spectrum disorders (ASD), auto-immune diseases, asthma, sterility and sexual dysfunction and others.' He concluded with the following particularly startling claim: 'The problem of sexual dysfunction is a huge problem in the USA, where males become impotent around the age of 24, at the prime of life.'[8]

On a visit to Harare University I saw the effect of Mugabe's anti-science policies at first hand. Young undergraduates studying genetics had been forced to salvage equipment from Harare's rubbish dump in order to learn the basics of biotechnology. With no Petri dishes or agar jelly, they had instead used washed jam-jars and home-made gels for plant tissue cultures. There were no functioning taps in the lab, so water had to be carried up two flights of stairs by bucket. Impressively, the enterprising students had managed to cultivate their own *Bt* crop embryos and even make progress on disease-resistant sweet potato, all with only intermittent electricity and occasional harassment from state officials. Their professor told me sadly that he was in the process of being hounded out of a job, both because he supported the use of biotechnology and because he was white.

Neighbouring Zambia, despite ideological differences with Mugabe, shares his blinkered approach to GMOs. No research or field trials are permitted in the country. As if to make the point in the most unsubtle way possible, in 2014 the Zambian authorities lit a choreographed bonfire of boxes of cornflakes, reportedly GMO-containing imports from South Africa and impounded from the shelves of local supermarkets.

A spokesman was quoted in press coverage saying: 'Health inspectors confiscated the cornflakes from a local leading chain-store after the Ministry of Local Government and the Attorney-General issued separate circulars urging Councils throughout the country to exercise their powers of seizure, disposal and destroy without compensation on any products containing Genetically Modified Organisms which are imported into the country.'[9]

Burning food in a country with 45 per cent rates of malnutrition[10] might seem like a calculated insult, but then Zambia has a troubled history in this regard. In 2002, in the midst of a severe drought that threatened Zambia and surrounding countries with famine, then-president Levy Mwanawasa banned GMO food aid on the grounds that he had been told it was 'poisonous'. As related by Harvard's Robert Paarlberg in his book *Starved for Science: How Biotechnology is Being Kept out of Africa*, Zambian authorities forced the World Food Programme 'to begin removing from Zambia the GM food aid supplies it had delivered earlier'. A *New York Times* article from the time quoted the Zambian agriculture minister saying: 'I have been told it is not safe.' Asked whether he thought GMO food aid was poisonous, he responded: 'What else would you call an allergy caused by a substance? That substance that the person reacts to is poisonous.'

According to the *New York Times*, the distribution of 14,000 tonnes of maize food aid was frozen by the Zambian president's edict. Ordinary Zambians, already suffering extreme hunger, were warned away from the food stored in warehouses. 'They have said that the food is not good for us, but we don't know ... they don't explain,' one citizen was quoted as saying. The article noted that families had been forced by hunger to go out into the bush and look for wild roots, and noted: 'With each passing day, the fates of millions of hungry people around Zambia grow more dire.'[11] According to Paarlberg, desperate Zambians had little time for the president's peculiar dogma, and in January 2003 'a mob of villagers in the town of Sizanongwe, 300 kilometers from the capital, overpowered an armed guard and looted several thousand bags of food aid before it could be removed.'[12]

Paarlberg alleges that Zambia's government was unduly influenced by strong lobbying from anti-GMO groups based in rich countries, including environmental groups such as Greenpeace. Not surprisingly, given how many lives were at stake, the Zambia famine episode remains highly controversial. In September 2002 Greenpeace rejected the accusation that it was somehow complicit in mass hunger, defending the Zambian president's decision to reject GM food aid on the basis that Mwanawasa 'knows that the future of his country's agricultural production is at stake'. According to Greenpeace: 'We say that as long as supplies of non-genetically engineered grain exist, nobody should be forced to eat genetically engineered (GE) grain against their will. If the choice really was between GE grain and starvation then clearly any food is the preferable option – but that's a false and cynical picture of the choice in this situation.' However, despite the urgency and sensitivity of the situation, the Greenpeace statement was hardly very nuanced. It described US-originated food aid as 'GMO-laced' and insisted that 'Zambia made a brave choice to preserve their agricultural heritage and its future'. It also insisted, against prevailing scientific wisdom even back in 2002, that 'GE foods are still an unknown quantity when it comes to health [and] safety', and that 'the weakened state of a malnourished population' might somehow make people more susceptible to the unknown health effects of GMOs.

Oddly, I have even been implicated myself in this debacle. One blogger in the *Telegraph* asked in 2010, 'What about the hundreds – perhaps thousands – of starving Zambians who died in the 2002 famine when, thanks to the misinformed campaigning of green activists like Lynas, the Zambian government refused to distribute US foreign aid packages of GM food?'[13] I had no direct involvement in the episode, so I presume the *Telegraph* blogger was referring to my anti-GMO activism in the preceding decade.

So did thousands of people die? According to the World Food Programme (WFP), while the drought initially left 'almost three million people in need of emergency food distributions', thanks to cash donations from other governments non-GM food was procured in time to avoid mass deaths from starvation. 'We were able to purchase food in the region and elsewhere, non-GM food, for distribution,' said a WFP spokesman to a television interviewer later. 'We had to go back to our drawing boards to replan our exercise,' Charles Mushitu of the Zambian Red Cross remembered. 'We started distributing beans from other neighbouring countries like Tanzania.' Thanks to this change of strategy, Mushitu claimed, 'We didn't record any single death arising out of hunger.'[14] However, there may well have been deaths that went unreported – not least because the Zambian government, after taking such a controversial stance, did not want to be blamed for its citizens starving. According to a *Guardian* report from October 2002, 'State radio and newspapers echo his [the president's] concern about GM and play down the food crisis. An MP who alleged in parliament that three constituents had starved to death was threatened with arrest.'

Whatever the truth, it seems clear that the situation could have been ameliorated more quickly had the Zambian government not stepped in with its GMO ban. This was not a 'brave' decision, as Greenpeace claimed, it was an ill-informed and reckless one that risked the lives and health of millions of men, women and children then suffering hunger and malnutrition. In one village visited by the *Guardian* reporter Rory Carroll, although 'there were no confirmed deaths from starvation', a resident told him that 'the children cry themselves to sleep from hunger. We go further into the bush to find nuts and berries, but it's not enough.' In a neighbouring village, Carroll was told: 'The children refused to go to school today. They said they were too weak to walk or concentrate. That never happened before, no matter how bad it got.' He noted that 'In the nearest town, Livingstone, several thousand tonnes of emergency relief maize sits in a warehouse with frosted windows on Industrial Road, untouched since arriving in July.' When he asked a villager about it, the man replied: 'The GM? Yes, the radio says it's poison.'[15]

There was another dimension to this injustice other than the pernicious influence of well-fed outsiders. This was to do with private-sector interests, in particular those of Zambia's luxury niche organic industry, which feared a loss of export markets to Europe. In 2002, as millions of Zambians faced the imminent danger of starvation, an organic exporting company 'received phone calls from British supermarkets explaining that exports of organic baby corn to the United Kingdom would be in jeopardy if food aid shipments containing GM maize were allowed into Zambia', according to Robert Paarlberg. 'The export markets demanded, that if there was any activity on GMO in Zambia then they wouldn't buy that particular crop,' confirmed Zambia National Farmers Union representative Songowayo Zyambo. Greenpeace too said as much in 2002: 'Africans fear genetic contamination because they can trade on the GE-free value of their grain and organically raised livestock. Profitable EU markets could evaporate if the slightest GE contamination rears its head.'[16] Let them eat organic baby corn – Marie Antoinette had nothing on this.

Of course the foreign-based NGOs fighting to keep GMOs out of Africa are not consciously aiming to worsen poverty and food insecurity. Their imagined aim is the opposite. Their means of doing so is by defending traditional lifestyles and heirloom crops, and promoting political objectives such as 'food sovereignty', defined by Food Sovereignty Ghana as 'the right of peoples to healthy and culturally appropriate food produced through ecologically sound and sustainable methods, and their right to define their own food and agriculture systems'. This sounds reasonable enough at first pass. But Paarlberg suggests that these political perspectives are themselves a European import, reflective of post-agricultural developed societies and unsuited to Africa. 'Low productivity in farming is the trap that is currently keeping most Africans poor,' he writes, echoing the farmers and scientists I spoke to in Uganda and Tanzania. 'European tastes regarding agricultural GMOs are not a good fit to the needs of Africa, given that two-thirds of all Africans are poor farmers in desperate need of new technologies to boost their crops' productivity.'

In food-stuffed North America and Europe, Paarlberg continues, 'New applications of agricultural science cease to be attractive, as they only seem harbingers of a still more heavily engineered and corporatized approach to crop and animal production. Quality-conscious citizens in rich countries at this point begin to desire less modern science and technology in their food production systems rather than more.' Lost farm productivity can easily be accommodated in higher prices when the cost of food is scarcely noticeable in most household budgets, especially when organic food already commands a hefty premium among the well-heeled. Sub-Saharan Africa, on the other hand, bears more resemblance to the Europe of the Middle Ages. Most work is done by hand with rudimentary tools; chemical fertilisers and irrigation are mostly non-existent, and famine is a constant threat. As Paarlberg explains, African farmers, themselves mostly women, 'have a food system that is traditional, local, non-industrial, and very slow. Using few purchased inputs, they are de facto organic. And as a consequence they remain poor and poorly fed.'[17]

Paarlberg's conclusion very much supports what I also found in Africa. He writes: 'Africa's rejection of genetically engineered crops today is far more western than it is African. Governments in Africa did not begin to get cold feet about GM crops until they saw activists and consumers in rich countries – particularly in Europe – rejecting the technology.'

In mid-February 2017, almost 20 years since I stood in a field of genetically modified maize in England and prepared to destroy it, I was back in Tanzania. The empty test site I had first visited in 2013 was now full of a flourishing and sturdy transgenic maize crop. This maize marked a milestone for the country as the first GMO test ever permitted; these were historic plants I saw rustling in a gentle tropical breeze. Wearing official blue overalls and baking under the intense sun, I looked around at the high fence and pondered at how I seemed to have come full circle. Two decades earlier I had destroyed maize looking very much like this, because it was also genetically engineered. This time I hadn't come to destroy the crop, I had come to help the scientists who were battling to defend it.

These scientists had won a significant battle the preceding year, succeeding in getting the 'strict liability' law relaxed so that they could legally conduct outdoor field trials of genetically modified crops. I was visiting this time not as an individual but as a visiting fellow of the recently launched Alliance for Science at Cornell University, and my mission was to work with the Tanzanian scientists so they could be as effective as possible in getting their story out to the world. The first crop to be planted was the maize I was now examining, part of the Water Efficient Maize for Africa (WEMA) project trialling drought-tolerant corn intended for the use of smallholder farmers in five African countries.[18] As well as Tanzania, WEMA maize was in development in Uganda, Kenya and Mozambique, and was already being distributed to farmers in South Africa. The maize plants were tall and strong-looking, easily the most vigorous I had seen so far in the whole country.

Tanzania's water efficient maize project seemed particularly appropriate because the entire East African region was suffering under a crippling drought, the worst in decades. According to the international Famine Early Warning System, Tanzania was then classed at risk of 'acute food insecurity'.[19] Half of the country's maize crop had been lost due to drought before the last harvest. As we drove from Morogoro to Dodoma to visit the trial site I could see that the road was lined with mile after mile of failed crops yellowing in their parched fields. It was a depressing sight, not least because I knew that it would be likely to mean hunger for the subsistence growers that depended on them.

Farmers I spoke to told a bleak story of rains that never came. 'The weather this year honestly is not like any other I have ever seen since I came to Dodoma; this is the most severe,' Regina Mwashilemo, one of these farmers, told me. She lived in the neighbouring village, Veyula, just a few dusty miles from the site of the GMO field trial. The proud owner of three emaciated cows, about a dozen chickens and a couple of goats, she supported five children and two grandchildren on just a few acres of land. 'It's not rained since November-December,' she complained. 'Now we're in February and still no rain completely.' Mwashilemo had heard of the WEMA trial down the road. For her the new drought-tolerant seeds couldn't come soon enough. She had already abandoned maize and gone back to growing the less valuable but drought-hardier crop, sorghum. 'Honestly, if I can get seeds for some good drought-resistant maize, I will go back to growing maize,' she said. 'I really need these WEMA seeds, and if they can give me just enough to plant, I will grow maize again so it can sustain me better.'

Mwashilemo was middle class compared to her near-neighbour Juma Chizuwa. A father of five, his family's desperate poverty was immediately betrayed by their tumble-down shack and ragged clothes. Chizuwa himself was stick-thin, as was his seven-year-old son Obama, optimistically named after the former US president. 'Honestly the climate is bad,' he told me when I asked how he was coping with the drought. 'Times are tough truly, only God will help us.' We wandered through the desiccated remains of his farm, which seemed like little more than a few wiry cassava plants adjacent to the desert-like remains of a dried-up river. I asked Chizuwa if he had heard of the drought-resistant seeds being trialled behind the high fence nearby. 'We're asking to get those seeds if possible,' he told me. And if he doesn't get help? 'I honestly can't tell how things will turn out.'

The fact that their country urgently needed their work to be successful had put a spring in the step of the Tanzanian scientists I was visiting. When I had first met Dr Alois Kullaya and Dr Nicholas Nyange four years earlier, they were frustrated and angry. This time they exuded a renewed sense of purpose, striding around the field trial site and tenderly examining the growing ears of the test maize crop. Dr Kullaya almost had to restrain himself as he told me how well it was going. 'From the general appearance, we think the genetically modified drought-tolerant hybrids are going to do better than the non-genetically modified,' he said jubilantly. 'We can say for sure when we have harvested and the results are out. But it looks very convincing.' Dr Nyange was equally keen. 'As I look to [this trial] there is a lot of hope that we can deliver this long-awaited drought-tolerant variety to benefit our poor resource farmers in our country,' he said smiling broadly.

But the scientists knew they must not get carried away. Dr Kullaya told me that at least another two years of field trials would be needed to test WEMA maize in different conditions in different parts of the country. The government would also need to relax 'strict liability' laws further to allow full commercialisation, because at the moment only scientific trials were permitted. And the anti-GMO groups were far from vanquished: a week after I left Tanzania one activist published a long screed in the country's leading newspaper the *Daily News*. 'GM technology is in effect turning Tanzania's 50 million plus people into guinea pigs, or perhaps even worse,' the writer claimed.[20] As usual, the article was highly misleading, and ended with an appeal to the country's president for a total ban on the WEMA trials and indeed all genetically modified crops. But the scientists had been smart: many media professionals and top policy-makers had already been invited to Dodoma and given tours of the WEMA crop. 'Seeing is believing' was their catchphrase, and as I saw for myself the drought-tolerant maize looked very good indeed.

But there was one final indignity to be endured. The confined field trial was still also subject to unjustified 'biosafety' rules. That was why I was required to wear the regulation blue overalls even in the hottest part of the day, and why all entrants into the area had to walk through a vat of disinfectant and past warning signs in capital letters. It also explained the high fence, and the fact that no genetically modified plant material was allowed to leave the site. Thus, after harvesting, which took place a fortnight after I left Tanzania, all the researchers lined up in front of a deep trench and set fire to several tonnes of perfectly edible food.

It must have been a very hard act to carry out. I was sent some photos later, of researchers standing grim-faced to the side of the trench as two tonnes of valuable maize was shovelled onto the fire.[21] I wondered whether Juma Chizuwa's hungry children might even have been able to see the smoke from their drought-stricken farm a few miles away. It struck me again as a symbol for how the rich world's fearmongering on GMOs has harmed the interests of the poor. After the burning, the trench was backfilled with earth, I suppose to protect the surrounding countryside from any drifting genetically modified ash.

Before leaving Tanzania after my return trip in 2017 there was someone I wanted to revisit. When I first met Grace Rehema in 2013, as detailed earlier in this chapter, she and her family were experiencing serious food shortages. Their cassava crop had failed due to viral outbreaks, and she didn't know where the next meal was coming from. Four years later, I wanted to find out what had happened. I knew that Grace would still not have had access to virus-resistant genetically modified cassava. Restrictive laws precluded it, and researchers in neighbouring Uganda had not yet applied for the crop to be released. So how would she be surviving without a healthy cassava crop? As we bumped down the same sandy track on the outskirts of Bagamoyo, I felt some trepidation.

Even with the help of a local Swahili-speaking scientist we had some trouble finding Grace. To assist us, we recruited a local teenager in a red T-shirt, and I got out my laptop to show him photos I took of Grace and her family four years earlier. He beamed with recognition: 'Over here!'

And there she was. Wearing a wrap-around orange skirt and pink top, Grace Rehema peered suspiciously out from a nearby hut. But once I showed her the photos on my laptop she burst out in delighted laughter. We gathered round to flick through the pictures, each one greeted with a shriek of happy recognition on Grace's part. She also explained how things had been in the intervening years. I was more than a little surprised to see a new hut on the site. It was made out of local materials, mud infilled between tree branches for walls, and with palm thatch for a roof. But it was new and even if cheaply built suggested that life had not been as disastrous for her as I had feared it might have been. Although her small cassava crop still looked bad, she had managed to diversify, she told me, selling mangoes by the roadside to make some cash to buy food. 'Where did the mangoes come from?' I asked. 'Look up!' she answered with another laugh. Overhead I saw the spreading branches of a huge tree, with ripening mangoes in their hundreds hanging in the midst of the shady green leaves.

I was surprised and pleased, as I had imagined back in 2013 that Grace Rehema had only one option, and that without her cassava crop she would simply sit there and starve. With my single-minded pursuit of a storyline that happened to fit my own interests at the time, I hadn't considered that she had multiple options, and many ways of being resilient. One of the scientist's most frequently heard refrains was that GMOs were 'not a silver bullet', and here I had first-hand experience of why this was true. Suddenly I saw Grace as a Tanzanian everywoman, someone who was everyone and everywhere. She was not just one thing, waiting to be described and defined by an agenda-focused outsider like me. She was a wife, a mother, a farmer and an entrepreneur. Yes, virus-resistant cassava, all other things remaining equal, might have helped her to improve her family's food security. It surely is not in her interests, nor the interests of people like her, to be denied an option that might improve her livelihood. But life is not monochromatic. She had adapted in different ways, and her kids were growing up well.

In fact I hardly recognised Grace's oldest girl. Just a small bareheaded child in my old photo, she was now a young woman of 13 years, wearing a smart yellow headscarf. Her mother told me proudly that her daughter Shida was doing well at school. I asked Shida, via a local scientist acting as translator, whether she would like to go to university and perhaps become a molecular biologist.

She turned, looked me straight in the eye, and answered in English: 'Yes'.

CHAPTER SEVEN

The Rise and Rise of the Anti-GMO Movement

'You know where the opposition to GMOs started? In my office. We started the whole opposition worldwide.'[1] So asserted Jeremy Rifkin, a globe-trotting American activist, futurist and writer, in a 2015 interview. This sounds like an inflated claim, but it is not without justice. Rifkin was undoubtedly the most important early individual involved in what later became a worldwide movement against genetic engineering, as I will show in this chapter. However, he was not the first to take seriously the potential negative impacts of biotechnology. Neither were Greenpeace, Friends of the Earth, nor any of the other environmental groups that later became prominent in the opposition campaigns to GMOs. In fact the earliest concerns were raised by the scientists themselves – those same pioneers who were in the vanguard of developing and deploying new hybrid DNA organisms.

The first time the issue arose was in the summer of 1971 when Paul Berg of Stanford University – who was later awarded the Nobel Prize for his work on recombinant DNA – proposed an experiment to combine a chromosome from the tumour-causing simian virus SV40 with the human intestinal bacterium *E. coli*. Berg postponed his work after receiving a worried phone call from Robert Pollack of the Cold Spring Harbor Laboratory. 'We're in a pre-Hiroshima situation,' Pollack later told *Science* magazine. 'It would be a real disaster if one of the agents now being handled in research should in fact be a real human cancer agent.' Pollack was not alone in his concern. 'This could recreate the conditions for an influenza pandemic like that of 1918,' warned Wallace Rowe of the National Institute of Allergy and Infectious Diseases (NIAID).[2]

The specific concern was that hybrid recombinant DNA might introduce novel pathogens with deadly consequences. As the co-discoverer of the DNA double helix, James Watson, later explained: 'Might some of the new genetic combinations that we would create in the test-tube rise up like the genie from Aladdin's lamp and multiply without control, eventually replacing preexisting plants and animals, if not man himself? If we assume that evolution can generate harmful variants, shouldn't we worry that creating novel combinations of DNA might have consequences orders of magnitude worse than natural disasters such as the lethal swine flu epidemic of 1918?'[3]

No one could say for sure what might happen, and the mere process of experimentation clearly raised risks. Researchers were especially vigilant because the dangers of working with pathogens, even using the highest standards of biosafety, were well known. 'Every microbiologist has inhaled or absorbed significant amounts of any organism he has worked with,' admitted one rather candid laboratory safety director to *Science*.[4] Accidents had sometimes been fatal: in 1967 a monkey agent, Marburg virus, infected 31 laboratory workers and others in Germany, resulting in seven deaths.

This was also a time of rapid social and cultural change, when many scientists were outraged by the Vietnam War and concerned that their academic institutions were complicit in supporting the so-called military-industrial complex by developing dual-purpose technologies. Science which started out as blue-skies research could potentially be turned to militaristic ends in different hands. How this intersected with genuine scientific enquiry, and whether this should ever be curtailed, was an open question. As Jonathan King, a biologist from the Massachusetts Institute of Technology, told a National Academy of Sciences meeting on recombinant DNA in 1977:[5] 'I was a graduate student at Caltech during the war years, where there were a lot of missile engineers. A number of us were concerned that these people were using their scientific skills to design devices to kill people. And we would raise questions sitting around the dormitory, and they would say you are interfering with our freedom of inquiry. What freedom of inquiry? You are making missiles. They would say we are not making missiles; we are studying the motion of an elongated projectile through a liquid medium, and if we cannot do that we cannot learn about it.'

Scientists were especially torn because they worried that research might be blocked across the board if their cautious warnings triggered a social panic. They were also very aware of how difficult most people found it to judge real risk.* Some were concerned that setting up new regulatory systems would hamper legitimate research. 'Even good intentions could easily lead to the creation of a self-sustaining and even growing bureaucratic monster which would discourage and delay very important research,' warned one oncologist, who was worried that this new bureaucracy could then delay or prevent breakthroughs such as in cancer treatments.[6] However, the researchers were aware that pressing ahead regardless would also be asking for trouble. 'If the public feels the scientific community is acting irresponsibly, there will be an immediate reaction and the freedom of research will be curtailed. If we don't exercise due caution we are heading for trouble,' warned one researcher on infectious disease in 1973.[7]

Many of the warnings were troubling because they came from scientists, even on political questions. 'I am a geneticist myself. I love genes. I love chromosomes. I make my living studying them,' said MIT's Jonathan King, a prominent member of the left-wing group Science for the People. Even so, he supported calls for a cessation of genetic engineering research, arguing: 'This is not a question of freedom of inquiry. This is a question of freedom of manufacture, of

---

* The most famous example was how people tended to be scared to death of nuclear power, but happily climbed into their cars, which were vastly more dangerous in terms of fatality rates.

modifying the environment, of modifying living organisms
... No one of us is saying don't accumulate knowledge.'[8] But
on recombinant DNA he wanted to call a halt. 'You may
argue that this is not a decision scientists should make,' read a
letter from a biolab director addressed to Paul Berg. 'I would
remind you of DDT and napalm – a few among products that
have not been used wisely.'[9]

Some of the researchers were clearly spooked by their own
work. 'The Berg experiment scares the pants off a lot of
people, including him,' admitted Rowe of NIAID.
Researchers were sufficiently worried, in fact, that they
declared a voluntary moratorium on all recombinant DNA
research using pathogenic viruses or bacteria in July 1974.
Signatories of a warning letter on 'Potential Biohazards of
Recombinant DNA Molecules' included luminaries like Paul
Berg and James Watson, and also Herbert Boyer and Stanley
Cohen, who had run the first experiments on recombinant
DNA. In order to thrash out a general position, 140 of the
world's top experts met at Asilomar, on the California coast,
eight months later in February 1975. Even at the time
participants had the sense that this was a historic occasion. As
Rolling Stone magazine reported, the molecular biologists
present were aware that 'they had clearly reached the edge of
an experimental precipice that may ultimately prove equal to
that faced by nuclear physicists in the years prior to the atomic
bomb'.[10] After sleepless nights and intense debate, the scientists
agreed to a degree of self-regulation, with strict biosafety
procedures attached to what seemed to be the highest-risk
experiments. To some, this was a moment for celebration –
the first time scientists had taken a degree of social
responsibility for their work before anything bad happened
rather than clearing up a mess afterwards when it was too
late. Others were less impressed at the strictness of the new
system: 'If such rules were to be applied to medicine the
hospitals would have to be closed,' complained one
correspondent in a letter to James Watson.[11]

As many researchers had feared, Asilomar was the
beginning rather than the end of the resulting process of

regulatory ratcheting. Governments were starting to take notice, as was the press, and the National Institutes of Health (NIH) responded by drafting regulations that were sufficiently stringent as to make some recombinant DNA research almost impossible. But this did not go far enough for the critics, who now included the environmental group Friends of the Earth. In a letter to *Science*, Francine Robinson Simring from the Friends of the Earth Committee on Genetics asked: 'What scientist would claim that complete laboratory containment is possible and that accident due to human fallibility and technical failures will not occur?'[12] Biochemist Liebe Cavalieri at Cornell University wrote a widely circulated *New York Times Magazine* piece in which he warned: 'Most problems of modern technology build up visibly and gradually and can be stopped before a critical stage is reached. Not so with genetically altered bacteria; a single unrecognized accident could contaminate the entire earth with an ineradicable and dangerous agent that might not reveal its presence until its deadly work was done.'[13] Simring later organised critics into a Committee for Responsible Genetics, which issued a bimonthly newsletter called *GeneWatch*.

However, just as the regulatory system was picking up steam, many in the scientific community were moving in the other direction. One of these was DNA pioneer James Watson. 'My position is that I don't regard recombinant DNA as a major or plausible public health hazard, and so I don't think that legislation is necessary,' he said.[14] Although Watson had signed the 1974 letter urging a voluntary moratorium on recombinant DNA, three years later he had seen enough evidence of its benign nature to change his mind. At the time of the letter, he explained, scientists had thought that recombinant DNA was entirely novel, that it had never existed in nature before and therefore might pose serious unknown risks. Now evidence had emerged showing that bacterial genes commonly pass into plants via plasmids to cause galls (this was the fruits of the work that was being undertaken by Jeff Schell, Marc Van Montagu and others around this time) and that natural genetic engineering probably happened in numerous other ways too. Rather than being the first time in three billion years that recombinant DNA had been created, Watson wrote, 'DNA transfer I think is probably fairly common in nature'.[15] It had also become known that viruses and bacteria exchanged genes spontaneously, and that laboratory pathogens tended to lose rather than gain virulence through 'domestication'. All in all, recombinant DNA had turned out in the light of new evidence to be less risky than originally feared. Watson concluded memorably: 'I'm drawing up a Whole Risk Catalog. Under "D" I have dogs, doctors, dioxin. Where do I put DNA? Very low.'[16]

Watson was not the only one to change his perspective. 'I have gradually come to the realization that the introduction of foreign DNA sequences into [bacteria] offers no danger whatsoever to any human being,' wrote University of Alabama scientist Roy Curtis to the director of the National Institutes of Health. 'The arrival at this conclusion has been somewhat painful and with reluctance since it is contrary to my past "feelings" about the biohazards of recombinant DNA research.'[17] But it was already too late to restrain the regulators. Under rising pressure from the public and environmental groups, concerned politicians led by Senator Edward Kennedy drafted a Bill to regulate recombinant DNA research, threatening drastic fines of $10,000 per day for breaches of the new code. Critics complained that the Bill resembled Soviet-era Lysenkoism (the pseudo-scientific ideology of Trofim Lysenko, endorsed by Stalin, that rejected conventional genetics and natural selection), and 137 scientists at one expert meeting wrote to Congress warning that Kennedy's draft legislation, if implemented, would 'inhibit severely the further development of this field of research'. A similar bill was proposed in the House of Representatives. After furious lobbying by scientific groups both bills were dropped, but there was no doubt that national-scale regulation was now in the offing. In future years, no fewer than three agencies of the federal government – the Food and Drug Administration, the Environmental

Protection Agency and the United States Department of Agriculture (USDA) – would be involved in the regulation of different GMOs, creating a cumbersome system with little scientific basis.

Jeremy Rifkin (whom we met earlier in this chapter) entered the frame, in trademark theatrical style, with a demonstration at a National Academy of Sciences meeting on 7 March 1977. Demonstrators from Rifkin's People's Business Commission held a banner saying: '"WE WILL CREATE THE PERFECT RACE" Adolf Hitler – 1933'. When Rifkin himself was invited to the microphone 'in the spirit of openness' by the presiding scientist, he proceeded to deliver a barnstorming speech accusing the scientists of being secretive, narrow-minded and politically naive. 'The one interesting thing about this Forum is that we are missing the central issue of why the subject is so important,' Rifkin declared.

> *We have heard for months both proponents and critics arguing that the real question here is safety. Is it safe or unsafe in the laboratories to conduct this experimentation? Do we need P1 laboratories or P4 laboratories? Do we need NIH voluntary guidelines or do we need involuntary regulations?*
>
> *My friends, the real issue is not whether the laboratory conditions are safe or unsafe, although obviously there is a problem with potential viruses and bacteria getting out of the laboratory and endangering the health and well-being of millions of people. But that is not the central issue. We could have legislation passed this spring by Congress for safety regulations, and it still would not detract from the central issue we are facing … The real issue here is the most important one that humankind has ever had to grapple with. You know it, and I know it. With the discovery of recombinant DNA scientists have unlocked the mystery of life itself. It is now only a matter of time – five years, fifteen years, twenty-five years, thirty years – until the biologists, some of whom are in this room, will*

*be able literally, through recombinant DNA research, to create
new plants, new strains of animals, and even genetically alter the
human being on this earth.*

Rifkin proceeded to quote from several leading scientists –
including James Watson – who he said were proposing human
cloning. He also predicted that religious groups would soon
join the tide of opposition, and declared it immoral that
corporations could own patents on life. 'That is all I have to
say. Let's open this conference up or close it down!' he
demanded.

If I had to put a date for the origin of the anti-GMO
movement, this would be the day. Prior to the demonstration
Rifkin and others had organised a new and diverse coalition
opposing recombinant DNA research, including two Nobel
Prize laureates and environmental groups such as Friends of
the Earth, the US Environmental Defense Fund, Natural
Resources Defense Council and Science for the People.
Friends of the Earth had now strengthened its earlier nuanced
position to one demanding 'an official moratorium on
recombinant DNA research', while the board of directors of
the Sierra Club announced that, pending further information
and discussion, 'the Sierra Club opposes the creation of
recombinant DNA for any purpose, save in a small number of
maximum containment labs operated or controlled directly
by the federal government'.[18]

It was ironic that at just the moment the scientific
community was beginning to realise that many experts'
initial fears about recombinant DNA had probably been
overblown, the environmental movement was solidifying
its position into one of implacable opposition. In 1977
James Watson was already looking back on Asilomar and
his own 1974 warning letter as 'a massive miscalculation in
which we cried wolf without having seen or even heard
one'.[19] But it was too late. Perhaps, in the spirit of Mary
Shelley's original novel, Frankenstein's monster was not so
much the science of DNA research as society's terrified
reaction to it.

One perhaps surprising defender of scientific research was the population biologist Paul R. Ehrlich, already a hero to many environmentalists for his championing of the issue of human overpopulation in his 1968 book *The Population Bomb*. Ehrlich wrote to Friends of the Earth in 1977 to urge it to drop its demand for a moratorium: 'In the case of recombinant DNA research,' he wrote, 'I think that the scientists have behaved admirably. Recognizing a possible serious hazard, they announced it to the public themselves and adopted voluntary restraints on their own research until the risks could be further examined.'[20] Ehrlich pointed out that not only had *Homo sapiens* been 'meddling with evolution' for a long time via selective breeding, but that as a population biologist he could see no reason why genetically engineered microbes produced in the laboratory should be able to out-compete natural bugs that are 'the highly specialised products of billions of years of evolution'.

Paul Ehrlich did not deny that science in the past had been misused – but he did not think that was reason enough to ban it. 'The results of virtually any pure scientific research have the potential for being turned against humanity,' he wrote, even his own apparently benign work on the coevolution of plants and butterflies. 'If recombinant DNA research is ended because it could be used for evil instead of good, then all of science will stand similarly indicted, and basic research may have to cease. If it makes that decision, humanity will have to be prepared to forgo the benefits of science, a cost that would be high indeed in an overpopulated world utterly dependent on sophisticated technology for any real hope of transitioning to a "sustainable society".'

But Friends of the Earth was not about to reconsider. Nor was Jeremy Rifkin. An American of impeccably radical credentials and a prolific writer possessed of an almost demonic energy, Rifkin eventually had, in the words of eminent social scientist Sheldon Krimsky, 'more impact on the media than any single

group or individual in the United States ... on genetics
policy'.[21] The food journalist Dan Charles agreed, crediting
Rifkin as being 'more responsible than anyone else for
awakening public fears on biotech' during his decades-long
campaign against genetic engineering. To cultural historian
Rachel Schurman, Rifkin was 'the individual who had the
greatest impact on expanding the anti-GE movement in the
United States (and to some extent globally)'. Starting out with
a flurry of books, articles and lawsuits beginning in the 1970s,
Rifkin also mentored and trained the subsequent generation
of US anti-GMO leaders from his small office just off Dupont
Circle in Washington DC.

Rifkin, like many of his compatriots in the early phase of
the anti-genetics movement, was an archetypal product of the
1960s counter-culture, raised in 1950s middle-class comfort
on the south-west side of Chicago. His mother Vivette made
tapes of books for visually impaired people, while his father
was a plastic bag manufacturer for industrial clients. Rifkin
was a highly promising student, graduating from the Wharton
School of the University of Pennsylvania with a degree in
economics and the school's Award of Merit in the tumultuous
year of 1967. Initially the student Jeremy showed no objection
to authority. According to a later profile in the alumnus
magazine, while at Wharton 'he became a cheerleader, class
president, a fraternity officer, and an economics whizz'. So far,
so conventional, but at the end of the 1966 academic year
something happened that would change Rifkin's life trajectory,
putting him on a journey of political radicalisation shared by
many students of his generation. 'One day there was a campus
demonstration,' Rifkin recalled. 'And I saw some football
players beating up kids. These [thugs] were my friends, the
same jocks I drank beer with at Smokey Joe's, and I thought,
"Wait a minute. Something's wrong here." I guess my
radicalisation began there.'[22]

The next day saw a very different Rifkin leading a freedom
of speech rally on campus to protest against the suppression of
dissent carried out by his former football friends. As the
counter-culture movement gained strength on campuses

across the US, Rifkin organised what he later claimed to be
the nation's first college sit-in, early the following year. Just a
few weeks before his graduation in 1967, Rifkin was on a
stage in front of 300 students at a rally protesting against the
Vietnam War. Now the sometime cheerleader was putting his
megaphone to another use. 'It is the responsibility of concerned
individuals to speak out and be counted,' the fraternity-
officer-turned-radical-activist urged the crowd.[23] Within a
year of graduation, he was helping to organise the 1968 March
on the Pentagon and had joined a self-described 'Citizens'
Commission of Inquiry (CCI) into US war crimes in
Vietnam'. The latter was given a boost as revelations about the
horrific My Lai massacre, where US troops murdered between
300 and 500 Vietnamese villagers, gradually leaked into the
public domain. As it became clear that the US military in
Vietnam was guilty not only of mass killings of civilians, but
also of attempting to cover them up, Rifkin and his colleagues
on the anti-war American political left began to gain influence
and credibility in their campaigns of opposition.

Under Rifkin's energetic and media-savvy leadership, the
CCI began a tour of the United States with Vietnam veterans.
The aim was to generate local press coverage by revealing
stories of other atrocities that had allegedly also been
committed by troops fighting in Vietnam. These Vietnam
vets claimed that My Lai was just the tip of the iceberg, and
that American war crimes were being committed on a larger
and more systematic scale than anyone realised. 'War
crimes ... are a matter of Pentagon policy,' Rifkin told the
New York Post in April 1970. This was a time of intense
national polarisation and soul-searching in the US. Was
America a benign superpower fighting to defend democracy
against the falling dominoes of Communism? Or was
American democracy nothing more than a sham, fooling the
working class at home and suppressing freedom in
the developing world? To New Left radicals like Rifkin the
shocking brutality of the shootings of unarmed students by
National Guardsmen at both Jackson State in Mississippi and
at Kent State in Ohio in 1970 reinforced the latter view.

One of the CCI Vietnam veterans, Michael Uhl, remembered Rifkin and wrote about this time in his memoirs. Jeremy was 'of Russian Jewish stock', Uhl wrote, and while at Tufts in 1969 his 'fascination with Hitler and horror of the Holocaust led him to research and write about eugenics and genocide' for his graduate thesis. Rifkin had been to Europe and visited the Nazi concentration camp at Dachau, 'and suddenly, in [his] mind's eye, [he] saw swastikas plastered all over American policy in Vietnam'.[24] According to Uhl, Rifkin's later 'relentless staying power' in opposing 'the genetically engineered adulteration of animals, vegetables or grains' was clearly linked to his 'youthful preoccupation with the fascist ideology of eugenics and its drive to engineer a master race'. Certainly these phrases and philosophical themes would reappear repeatedly in Rifkin's later writings about genetics, as they would in the wider anti-genetic engineering movement that he helped to establish.

The late 1960s New Left was very different to the Old Left. Not for Rifkin endless late-evening discussions about Trotsky's writings on dialectical and historical materialism. From the outset he was a pragmatist, measuring impact in newspaper column inches rather than degrees of ideological purity. While the Old Left was primarily concerned with class relations within nation states, the New Left was much more interested in the transnational power of big business. As Vietnam faded from centre stage in the 1970s, Rifkin launched a Peoples Bicentennial Commission (PBC), later the Peoples Business Commission, as an anti-corporate alternative to the official 200th birthday celebrations of the 1776 founding of the United States. One of its first headline-grabbing actions in 1973 was to dump fake oil drums into Boston harbour as a symbolic oil-age echo of the 1773 Boston Tea Party. A later attention-grabbing idea was to offer a $25,000 reward to any company secretary who would pass on insider information of corporate malfeasance to his PBC. The proffered cash would be payable 'for concrete information that leads directly to the arrest, prosecution, conviction and imprisonment of a chief executive officer of one of America's Fortune 500 corporations for criminal activity relating to corporate operations'.[25] Rifkin sent a similar letter to the wives of the same corporate CEOs at their home addresses, with an attached tape recording promising cash in return for whistleblowing on their high-flying husbands.

The themes in Rifkin's 1976 letter to the corporate wives sound eerily familiar. He wrote: 'Today, 200 giant corporations already own over two-thirds of the manufacturing assets of the country. Heading up these corporate empires are a small group of nameless, faceless men who have amassed enough power to virtually dominate American life ... Your husband is part of this small privileged business elite. That puts a special responsibility on you and your family to speak up against corporate policies that result in price-fixing, induced unemployment, environmental destruction, excessive profiteering, unfair distribution of wealth, and other abuses.'[26]

Rifkin's gambit did not pay off. All his Peoples Bicentennial Commission received in return was 'expensive, grainy engraved stationery with expletives' rather than the hoped-for salacious corporate gossip.[27] But Rifkin got his name in the *New York Times*, the *Wall Street Journal* and many other local and regional newspapers up and down the land, which was probably more the point anyway. His career as a media provocateur was beginning to take off.

Rifkin's success with the media was only partly down to his well-honed sound bites and his legendary ability to work a room. Rifkin also presented himself to journalists as a public intellectual, someone who had considered the deeper philosophical questions about technology and progress that eluded the blinkered scientists shuttered in their academic silos. Rifkin began to build the case for his opposition to genetic engineering in 1977, the same year as his National Academy of Sciences genetic engineering protest, with the publication of his book *Who Should Play God* (co-authored with Ted Howard).

Who Should Play God was dedicated to *Brave New World* author Aldous Huxley. 'He foresaw', Rifkin wrote ominously.

The very first page, under the headlined question WHAT KIND OF FUTURE IS THIS? made some dramatic statements, including allegations that scientists were intent on producing 'live carbon-copy clones of *you* in less than ten years' (emphasis in original), that 'normal sexual reproduction' might be 'totally replaced' by artificial reproduction 'within fifty years', and perhaps most surprisingly of all, that humans might be genetically engineered so they could digest hay like cows and photosynthesise via their skin.

Unless stopped in its tracks, genetic engineering would, Rifkin insisted, be 'a form of annihilation every bit as deadly as nuclear holocaust'.[28] Moreover, as implied repeatedly in the cover notes and endlessly repeated in speeches, he considered the technology to be inseparable from eugenics. Previous to recombinant DNA, Rifkin claimed, the 'symbiotic relationship between genetic engineering and social policies and visions ... found its high-water mark in the genetic policies of Hitler's Third Reich between 1932 and 1945.'[29] Accordingly, Rifkin foresaw 'the emergence of a kind of corporate Brave New World'. While this would admittedly be 'a much less dramatic approach to the ultimate enslavement of the human species ... the results are no less terrifying than if they had been ruthlessly imposed by some mad political dictator' like Hitler.[30]

Rifkin ended the book with the following prediction: 'If genetic engineering were allowed to continue at its current pace, the *Homo sapiens* species would experience no more than five or six more generations before being irreversibly replaced by a new, artificially engineered organism. Though this new species would include some of our characteristics, it would in many ways be as different from us as we are from our closest relatives, the primates.'[31] Opposing genetic engineering had therefore become for Rifkin a sacred duty akin to stopping the Third Reich. There could consequently be no co-existence between the science of genetics and basic humanitarian values. One or other of them must be eradicated. His mission was therefore necessary not just to forestall the rise of a new fascism and eugenics, but to save the entire human species from treading an otherwise inevitable path to extinction.

By 1985 Jeremy Rifkin was at the peak of his influence. What was once the Peoples Bicentennial Commision, then the Peoples Business Commission, had now been grandly renamed the Foundation for Economic Trends, occupying an expanded office in Washington DC. An interview published in the November 1985 edition of *Mother Jones* magazine carried a photograph of Rifkin in carefully posed profile, wearing his trademark beige slacks and seated in an office lounge chair. One finger was held thoughtfully to his moustachioed upper lip, the other hand clutched his ever-ready pen and open notepad. The accompanying interview, penned by *New York Times* writer Keith Schneider, portrayed Rifkin as an island of authoritative calm surrounded by intense and frenetic office activity. Secretaries would buzz him constantly with the latest media request: 'Jeremy – CBS News, line one.' 'Jeremy, Christine Russell from the *Post* wants to meet here after lunch. Should I tell her OK?' 'Jeremy, the House Energy and Commerce Committee, line two.'

The *Mother Jones* piece caught Rifkin in his anti-biotech heyday, where he was, in Schneider's words, 'the nation's most prominent critic of genetic engineering', whose special gift had been in 'gaining attention for a message that has put him at the bulls-eye of a global debate'. In only two years, Schneider wrote, Rifkin had 'emerged to command the epicenter of the moral and scientific discussions surrounding the development of genetic engineering'. As a second article related, Rifkin's office was a media production line. 'Two middle-aged women cut stories from the day's newspapers, articles that either quote Rifkin directly or make references to his multitude of pet causes. The stories are carefully copied and [...] distributed to interested parties. It is clear that without the media, there is no Rifkin.'[32] With his implacable opposition to genetic engineering and his talent for spreading his ideas far and wide, Rifkin was streets ahead of his opponents in the scientific community. 'The majority of scientists believe that Rifkin's warnings [about genetic engineering] are nothing more than the hyperbolic ventilations of a self-serving kook.' Schneider reported. But this same 'majority of scientists' were quickly discovering that they were virtually powerless to stop him.

Rifkin built his reputation not just in the media but through a punishing schedule of stump speeches, during which he projected a persuasive and charismatic persona perhaps more akin to a faith healer than an environmentalist. As one profile related: 'He sits on the edge of a table, loosens his tie, unbuttons his top shirt button, and carefully rolls up his sleeves. He picks up a microphone and begins to walk around the audience, making eye contact with nearly everyone in the room. He sips Evian water, cracks jokes, and glides from subject to subject. His standing-room-only audience seems transfixed.'[33]

The *Washington Post*, in a detailed 1988 profile focusing on Rifkin's anti-genetic engineering proselytising trip to Italy in the same year, discussed the back-room mechanics of what its reporter dubbed 'The Speech': 'It is a little as if Rifkin were a sophisticated photocopier; he can enlarge or reduce The Speech to whatever size is convenient.' And how was The Speech – of whatever size – delivered? 'He moves. He walks, he stalks, he points the portentous finger.' Rifkin's Italian hosts were somewhat perplexed by this brash American in their midst: 'Jeremy, he has a way of speaking to people like a preacher, or something similar, like a prophet,' confided one of them to the *Post*'s reporter. But Rifkin invariably succeeded: by the end of his trip he had persuaded the newly elected Italian Green Party to adopt opposition to genetic engineering as a central tenet of its policy platform. The *Washington Post* reporter who accompanied Rifkin over several days of frenetic travel reached this rather ambivalent conclusion: 'Jeremy himself is engaging, fun company. But he is also an alarmist and an absolutist, with little or no trust in humans to think for themselves. One shudders for a world in which Rifkin is king.'

Rifkin's tendency towards all-or-nothing Manichean thinking was evident in all his speeches, particularly on the subject of genetic engineering. As he expounded to students in an anti-globalisation 'teach-in' in 2001 at Hunter College, a New York liberal graduate school: 'Let me say to all of you. If we allow the great gene pool and the proteins it codes for to be enclosed as either political property owned by governments or intellectual property owned by life science companies, I guarantee you parents, your children and grandchildren will have gene wars in the 21st and 22nd century!' The theme of eugenics was a lifelong preoccupation. 'This new eugenics is friendly. It's banal. It's commercial. It's market-driven. Don't you all want a healthy baby?' Well yes, of course. 'But the problem is it fundamentally changes the parent-child bond, that's why it's a new eugenics. The parent becomes the architect, the child becomes the ultimate shopping experience in this post-modern world.'

This was reminiscent of a speech he had given decades earlier, at Gettysburg College Student Union in 1979: 'Genetic research is going to bring us one step closer to genetic engineering. That's where they tell us to produce ideal children and the last time that happened they had blue eyes, blond hair, and Aryan genes.' To the young Rifkin there were no shades of grey: 'You cannot be a genetic engineer without being a eugenicist,' he insisted.[34] Rifkin's demand was always the same, issued as a ringing declaration designed to be followed with thunderous applause. 'There should be a strict global moratorium, no release of any genetically modified organisms into the environment of this planet. Clear and simple.'

Jeremy Rifkin didn't just talk about banning genetics; he acted. 'If you look back, my lawyers brought the first case against GMOs,' he said recently. 'We stopped the first GMO, Ice-minus, from being released into the environment with a federal court decision in the US, which started this whole debate. Then we opposed patents on life at the Supreme Court. We've spent twenty to thirty years on this thing.'[35] During the 1980s and early 1990s Rifkin's Foundation on

Economic Trends issued a fusillade of lawsuits that successfully gummed up the progress of biotechnology in America. 'He has an uncanny ability to pinpoint weaknesses in our review procedures, and he is able to see much larger trouble spots ahead,' one EPA lawyer, having found himself on the receiving end of yet another lawsuit, told the *New York Times* with grudging admiration in 1986.[36] 'You hear all the time that this guy is nothing more than a nuisance. It's not true. He's a factor in almost every facet of biotechnology right now.' Just three years after the first recombinant DNA product, human insulin for diabetics, came to market, the *New York Times* was already writing about biotechnology as a 'stalled revolution'. Rifkin's Foundation by then included two full-time lawyers, more than sufficient to affect national policy.

Rifkin enjoyed his first taste of success in May 1984 when he managed to convince a Washington DC court to issue an injunction against the test spraying of genetically engineered bacteria by the University of California on a small plot of strawberries. The lawsuit was procedural – Rifkin argued that an appropriate environmental impact statement had not been prepared – but it did its job well. This was to be the first deliberate release into the environment of GE bacteria, so it was an important litmus test for both sides. These bacteria had had a gene removed in the lab to eliminate their ability to nucleate ice crystals on plant leaves, with the aim of protecting cold-vulnerable crops like strawberries and potatoes against frost damage. The court's decision, which largely went Rifkin's way, was a serious embarrassment to both the University of California and the National Institutes of Health, the agency formally tasked with setting and monitoring guidelines on the use of recombinant DNA technology. 'The consequences of dispersion of genetically altered organisms are uncertain,' the judges affirmed, and the NIH's procedures 'utterly fail[ed]' to meet the requisite standard.

A couple of years later, in May 1987, a private company called Advanced Genetics Systems resumed the stalled experiment, aiming to commercialise the bacteria as an agricultural spray with the trade name 'Frostban'. Although this time Rifkin's anticipated lawsuit failed to halt the experiment, EarthFirst! activists destroyed one of the test sites in the night. To my knowledge this was the world's first anti-GMO crop trashing action. As a BBC report later put it: 'The world's first trial site attracted the world's first field trasher.'[37] A parallel experiment with potatoes was also targeted by night-time activists. 'When I first heard that a company in Berkley was planning to release these bacteria Frostban in my community, I literally felt a knife go into me,' one of the EarthFirst!ers told the BBC. 'Here once again, for a buck, science, technology and corporations were going to invade my body with new bacteria that hadn't existed on the planet before. It had already been invaded by smog, by radiation, by toxic chemicals in my food, and I just wasn't going to take it anymore.' Rifkin was now making converts among the radical fringe, adding an extra string to his bow: the possibility of direct action when lawsuits and media campaigns failed.

The scientists admittedly didn't help their own cause. Health regulations meant that technicians had to spray the remaining strawberries with GE bacteria while dressed in scary-looking moon suits. As the *New York Times* reported later, 'Photographs of scientists in regulation protective gear – spacesuits with respirators – were broadcast around the world, generating widespread alarm,'[38] even though the photographers snapping the pictures were only a few feet away and themselves were wearing no protection. It was a classic example of how precautionary regulations aimed at reassuring an anxious public can have exactly the opposite effect.

Although the product passed its scientific tests, successfully inhibiting ice formation and protecting the plants, Frostban was already so controversial that it never made it to market. The nascent biotech industry was being forced to absorb some bitter lessons. Even if a proposed product could be tested, progress might be stalled for many years by lawsuits, hugely raising costs. Even after court approval, activists might then destroy the experiment physically. And even if a GE product managed to jump both these hoops, it might never make it to market because of continued opposition and negative media attention resulting in boycotts or worse. The result was a chill that spread throughout the biotechnology community. Now plant science using genetic engineering had a new and potentially highly expensive risk attached to it, the risk of public rejection.

This danger was well illustrated with the ill-fated launch five years later, in May 1994, of the world's first genetically engineered food product. Originally designed for the fresh foods market, Calgene's 'Flavr Savr' tomato was engineered to have a longer shelf-life and improved flavour via a reversed gene inhibiting the expression of the ripening enzyme polygalacturonase. The aim was reasonable enough. Rather than mass-market tasteless supermarket tomatoes, picked green and ripened artificially, consumers would be able to enjoy tomatoes picked when naturally ripe on the vine. Contemporary TV news reports show consumers cautious but generally positive. The taste was reportedly better, achieving the desired summer home-grown flavour. But Rifkin was having none of it. 'We're determined that genetic-engineered foods will not reach the market here or in Europe,' he vowed. Via a new offshoot from his Foundation, the Pure Food Campaign, Rifkin launched a two-pronged effort to demonise the new tomato in the public mind and to challenge it in the courts. 'For the taste of a tomato,' he asked rhetorically, 'you're going to risk a threat to your health or your children's health?'[39] Archive television news footage shows Rifkin declaring: 'It may be benign. But it may be toxic. Our position is: better safe than sorry.'[40]

Although his court challenges didn't succeed, Rifkin's threat of a nationwide boycott convinced Campbell's not to put the tomato in its canned soups. The Pure Food Campaign mobilised thousands of restaurant chefs who made a commitment not to serve the offending tomato, and mailed 'educational materials' to over 100,000 schoolteachers.[41]

Calgene's launch into the fresh tomato market, which was hampered by logistical problems as well as opposition from campaigners, didn't last long. In Europe a successor to the Flavr Savr briefly appeared on supermarket shelves as canned tomato paste, prominently labelled as 'genetically engineered'. Although it sold well, anti-GMO campaigners convinced the supermarkets to remove it and by 1999 the Flavr Savr had gone.[42] Calgene itself was sold to Monsanto, not because Monsanto wanted to get into the fresh tomato business, but in order to assume control of Calgene's valuable patent portfolio. The Flavr Savr was unceremoniously dumped. The first genetically engineered food product in the world had disappeared into history.

Rifkin's most enduring legacy, in my view, was through the other activists he gathered around him and inspired. Many of today's most influential anti-GMO leaders learned their craft alongside him, and continued to spread the message even as Rifkin himself moved on to different subjects in later years. One of these influential associates was Andrew Kimbrell, a trained lawyer who steered many of Rifkin's most important anti-GE lawsuits through the courts. Kimbrell later went on to establish the Center for Food Safety, which took over from Rifkin's Foundation on Economic Trends as both the intellectual and financial powerhouse of the anti-GMO movement from the 2000s to the present day.

Kimbrell's opposition to genetic engineering was just as implacable as Rifkin's. Whether plant or animal, public sector or private sector, he sought a total ban on biotech in food and agriculture, and a wholesale shift to worldwide organic farming. Another protégé of Rifkin's was Ronnie Cummins, a beret-sporting radical activist who took Rifkin's Pure Food Campaign and relaunched it as the Organic Consumers Association (OCA). In recent years OCA has funded the pro-labelling group US Right to Know, which has targeted numerous university biotechnology scientists with 'Freedom of Information' requests resulting in the release of thousands of emails and associated media controversies.[43] OCA also takes a radical anti-science position on conventional medicine: it has published material claiming child vaccines cause autism,[44] that homeopathy can guard against flu, and promoting intravenous vitamin C for the purpose of curing Ebola.[45] 'It is important to know how to protect your children and yourself with homeopathic and natural alternatives to vaccines to build your natural immunity to the swine flu,' states one page on OCA's website.[46]

Rifkin's single most important long-term contribution to spreading the campaign, however, was surely his recruitment of Greenpeace to the anti-GMO cause. Like many momentous events, this happened almost by accident when Rifkin had an impromptu meeting with a German activist called Benny Haerlin in 1986. Haerlin had impeccably radical roots: he had been an active member of Berlin's squatter movement in the early 1980s and had even been imprisoned on charges of terrorism after his magazine *Radikal* published manifestos by the Revolutionary Cells, an underground anarchist group responsible for bombings and plane hijackings in the previous decade. On his release from prison, Haerlin was recruited by the increasingly successful German Green Party and became a Member of the European Parliament.[47] It was as a Green MEP that Haerlin first crossed paths with Rifkin. On a visit to the United States in 1986 Haerlin met Linda Bullard, an idealistic employee of Rifkin's Foundation on Economic Trends. Bullard, later president of the International Federation of Organic Movements, brought Rifkin and Haerlin together. Rifkin's pitch converted Haerlin to the anti-biotech cause and he became the leading voice against genetic engineering in the European Parliament.[48]

The arrival of the first GMO foodstuffs into Europe in 1996 proved to be a turning point that catapulted the anti-genetic engineering movement into the global big-time. As Schurman and Munro write: 'Perhaps the person to recognize the propitiousness of the moment most explicitly was Benny Haerlin, the former German Green MEP who had been working to stop biotechnology for a decade. In the summer of 1996, Haerlin was coordinating an anti-toxics campaign for Greenpeace International when he received a call from an

executive at an upscale German supermarket chain called
Tengelmann's.'⁴⁹ The supermarket boss told Haerlin that he
knew the first shipment of GM foods was to arrive into
Europe later that year, and he wanted to know whether
Greenpeace would have a problem with that. At the time
Greenpeace did not have an anti-GMO food campaign in
place, but – thinking on his feet – Haerlin told his caller that
Greenpeace would indeed have a problem with the new
foodstuffs. He then hung up and immediately set about
organising a campaign.

As Dan Charles recounts:

> *Haerlin convinced Greenpeace that this was a crucial moment to*
> *push the GMO issue and persuaded the organization to assign*
> *fifteen full-time organizers to the issue. When the ships carrying*
> *the crops arrived in European ports, Greenpeace activists were*
> *ready and waiting … The activists swarmed the ships, prevented*
> *them temporarily from docking, and unfurled banners calling for*
> *a ban on the import of genetically engineered food.*⁵⁰

Friends of the Earth also established a major international
campaign against GM foods starting in 1996.

According to Dan Charles, despite the Greenpeace actions
Monsanto initially thought it had scored a victory. 'Across
most of Europe few people seemed to care. Even in Germany,
Denmark, and the Netherlands, formerly the strongholds of
opposition to biotechnology, Greenpeace wasn't able to stir
up much public reaction … There was even less reaction to
the south in Italy, Spain and France.'⁵¹ Greenpeace was also
surprised at how popular the new Roundup Ready soybeans
proved to be with American farmers. With as much as a third
to a half of the arriving shipments now comprising genetically
engineered soybeans, one Monsanto staffer recalled later:
'Bob Shapiro's [Monsanto's CEO] reaction was, the battle was
over and we'd won.' But if Shapiro was looking at Germany,
France or the Netherlands to gauge the public reaction to
biotechnology, he was focusing on the wrong target. The
whirlwind of opposition was about to emerge not from Paris

or Brussels but from London. As I detailed in the first chapter, it was the UK that became the epicentre of the worldwide movement against GMOs.

While back in 1996, attempts to impose GE labelling in Europe had failed due to lack of interest,[52] four years later – after the UK anti-GMO campaign had gone global – the EU was frightened into operating a self-imposed moratorium on all GM crop approvals. When the labelling issue came back before the European Parliament the following year, the assembly passed what was then called 'the toughest GMO legislation in the world' by an overwhelming 338 votes to 52.[53] In France the farmer-activist José Bové, whose Asterix-style moustache seemed to symbolise his populist Gallic demand for traditional food, destroyed GE rice in Montpellier in 1999.[54] France went on to become one of the most determined GMO-free countries in the whole of Europe. In India, meanwhile, a radical farmers' movement in Karnataka state took an advertised 'Operation Cremate Monsanto' action literally, on 28 November 1998 burning a test field of Monsanto GM pest-resistant cotton to the ground.[55] Three more fields were burned over the subsequent month. Arsonists also struck in Italy, where in April 2001 a Monsanto seed depot was burned down, with 'Monsanto killer: No GMOs' spray-painted on a wall.[56]

The campaigns, and their widespread media support, led to a massive decline in public support for biotechnology. In Britain and France, the percentage of the population opposed to GM foods rose by 20 points between 1996 and 1999.[57] Although the impacts were varied, opposition rose in every European country. In total only a fifth of Western Europeans remained supportive of GM foods, a radical turnaround from just a few years earlier when most people were either broadly in favour or didn't care either way. In response to rising fears, the European Union set up a cumbersome regulatory process that was 'process-based' rather than 'product-based': in other words, it singled out GMOs based on the process of molecular breeding rather than considering any meaningful difference in the resulting foodstuffs. Plant breeders could continue to use traditional random methods like mutagenesis, which was specifically excluded from the new regulations even though it might have significant effects on the biochemistry of target crops. If molecular biologists used the more precise technique of genetic engineering, however, they were required to submit 13 separate categories of technical data, often comprising hundreds of pages and costing tens of millions of dollars to compile.

The regulatory regime also allowed for political votes from member states: applications were held up for years, even decades, as arguments raged in the EU's Council of Ministers. Activists lobbied countries that were sceptical about GMOs, using the complicated EU approval systems to slow down and ultimately block these crops. From 1998 until the present day the approvals process has remained virtually deadlocked. Reluctant as ever to take a decision in the face of likely public hostility, European countries kept on asking for more data each time the scientists at the European Food Safety Authority recommended a GMO application for approval. And so it went on, with applications being batted around in Brussels like a ball in a squash court (except a good deal more slowly). Nearly 20 years later, in 2017, there has not been a single approval of a genetically modified crop for domestic cultivation. Thanks to the sudden rise of the anti-GMO movement, Europe slammed the doors on biotechnology in 1998 and has kept them shut ever since.

As the debate over biotechnology intensified, bigger money began to come into play. In 1997 US-based grant-makers established the Funders' Working Group on Biotechnology, which over the following three years allocated between 2 and 3 million dollars to anti-GMO activism, helping to support new coalitions that brought together numerous different groups. This amount was still dwarfed by the resources available to biotech industry lobbyists: Novartis, Monsanto and the other big beasts of the industry pooled resources to set up the Council for Biotechnology Information in 1999, with an annual budget of between 30 and 50 million dollars, more than ten times what was available to activist groups.[58] The companies also already had the Biotechnology Industry Organisation and the Grocery Manufacturers Association fighting their corner, both well-staffed and well-connected Washington lobby groups.

On the other side, the Rockefeller oil fortune heiress Mary Rockefeller Morgan began supporting the rising anti-biotechnology movement from the late 1990s onwards. Donors and activists were concerned that the European campaign against GMOs, which reached its apogee in the late 1990s, was passing the United States by. 'Activists are doing good things,' the 2001 document[59] quotes Chris Desser, coordinator of the Funders Working Group on Biotechnology, as saying, 'but they don't have the money to get the word out.' In an effort to change that, Mary Rockefeller Morgan penned a letter to other members of the family and launched a new 'Genetically Modified Foods Collaboration' in 1998 with the participation of six other Rockefeller inheritors.

One of the beneficiaries of this fundraising effort was Andrew Kimbrell's Center for Food Safety (CFS), helping it to grow rapidly and become the mainstay of the US anti-GMO movement over the following two decades. Half a million dollars was donated to CFS between 2002 and 2011 by Rockefeller Philanthropy Advisors, while $3 million more came from the Rockefeller-backed Cornerstone Campaign. According to its tax returns, annual revenues in 2014 crossed the $5 million mark, and CFS today has over 30 employees and offices in San Francisco, Portland and Honolulu as well as its headquarters in Washington DC. All told, CFS raised a total of $16 million cumulatively over the five years from 2009 to 2014, spending upwards of a third of a million dollars annually on political lobbying activities.

Another early supporter of anti-GMO causes was Doug Tompkins, the late millionaire outdoor clothing magnate turned deep ecologist. Tompkins' initial contribution in 1998 was to bankroll – to the tune of half a million dollars over the following four years – Kimbrell's second book, *Fatal Harvest: The Tragedy of Industrial Agriculture*, which brought together contributions from critics of modern agriculture ranging from the poet-farmer Wendell Berry to anti-globalisation activist Helena Norberg-Hodge and the Indian anti-biotech campaigner Vandana Shiva. As an early devotee of deep ecology, Tompkins was highly critical (in the words of the Foundation for Deep Ecology that he established) of 'industrial culture, whose development models construe the Earth only as raw materials to be used to satisfy consumption and production – to meet not only vital needs but inflated desires whose satisfaction requires more and more consumption.'[60] The Foundation's book *Work in Progress* adds, 'Modern society is enamoured with technology and accepts new technologies uncritically. In time it often becomes apparent that it would have been wiser to avoid developing or spreading that technology. It is now easy to imagine how the world might have been better off without nuclear technology, the Green Revolution in agriculture, gunpowder, television, internal combustion engines, and so on.'[61]

Kimbrell's Center for Food Safety received $1,670,000 from Doug Tompkins' Foundation for Deep Ecology between 1996 and 2003, making the Center a beneficiary second only to his former boss Jeremy Rifkin's own Foundation on Economic Trends, which received $1,812,000 in the same period.[62] Tompkins also contributed over a million dollars for a fusillade of full-page ads that ran in the *New York Times* during 1999, variously warning about the perils of globalisation, criticising advanced technology and denouncing the 'genetic roulette' of crop biotechnology.[*] Tompkins later put the majority of his fortune into buying and conserving a huge area of Patagonian wilderness, now recognised by Chile as a national park and called Parque Pumalin.

---

[*] This effort was called the Turning Point project, and the ads were reportedly penned by Kimbrell and Jerry Mander, a former ad-man.

The story of how much money went where is important because of the competing David and Goliath narratives of the different sides of the GMO debate. While activists bemoan the huge financial heft and political power available to corporations like Monsanto and associated private-sector lobbying groups, pro-biotech advocates contrast this with the hundreds of millions of dollars that now pour into the wider environmental movement. With big companies like Whole Foods and the wider organic industry joining the fight, perhaps the true picture today is more Goliath versus Goliath than the grassroots struggle I remember from the 1990s. Greenpeace reported expenditures totalling 321 million euros in 2015,[63] although only a small proportion of this would have been spent on anti-GMO campaigning. The organic foods sector now has a global turnover of more than $60 billion per year, and organic marketing makes great play of its non-GMO status.[64] One analysis produced by a pro-biotech consultant concluded: 'Some 300 formal and informal organizations with combined annual expenditures of $2.4 billion in annual revenues are involved in anti-GMO advocacy in North America.'[65]

Friends of the Earth, on the anti-biotech side of the argument, produced a report in 2015 entitled *Spinning Food: How food industry front groups and covert communications are shaping the story of food*.[66] This concluded that: 'The industrial food and agricultural sector spent hundreds of millions of dollars from 2009 to 2013 on communications efforts to spin the media, drive consumer behavior and advance its policy agenda.' This included '$126 million spent by 14 food industry front groups' such as the US Farmers and Ranchers Alliance and the Coalition for Safe and Affordable Food, the latter established by the Grocery Manufacturers Association to fight GMO labelling. The Friends of the Earth figure also included 'more than $600 million spent by four major trade associations – CropLife America, BIO, Grocery Manufacturers Association, and the American Meat Institute – that promote and defend the agendas of pesticide, biotech and conventional food corporations'. The report added that 'in 2013 Monsanto alone spent $95 million on marketing'.

I offer no definitive conclusion here other than to acknowledge that I too have a financial stake in the issue. While I have never received financial compensation from Monsanto, or indeed any other biotech company, I worked with the Alliance for Science at Cornell University for three years until 2017, which was established in 2014 with an initial $5.6 million grant from the Bill and Melinda Gates Foundation (renewed with a subsequent $6.4 million grant in 2017). I have also given paid keynote speeches to farm groups in the US and Canada – the latter usually during the freezing but agriculturally quiet months of January and February in out-of-the-way places like Portage la Prairie in Manitoba. I particularly remember the latter event, which I think was a potato growers' conference, because it was the only time I have experienced an outdoor temperature of -40°C, during a brief walk around the block which almost cost me my nose and earlobes. Everyone needs money; the important thing is transparency, so that conflicts of interest are not obscured and everyone can see who is backing whom.*

In summary, the anti-GMO movement has surely achieved one of the most stunning reversals in the history of citizen activism. Despite the early fears of its scientific pioneers in the 1970s, by the mid-1990s it looked to all the world as if genetically engineered crops would be the start of a new revolution that would transform world agriculture. Monsanto and the other biotech behemoths confidently expected that genetically engineered wheat, rice, potatoes – indeed all the world's staple food crops – would be in farms and on dinner tables in just a few years. It did not happen. Instead the impact of plant biotechnology was limited to just a few crops and – with the exception of *Bt* cotton, canola in Australia and some very small-scale pro-poor GMOs such as *Bt* brinjal and maize

---

*You will find the Cornell Alliance for Science's sources of funding listed on its website.

in developing countries – big agribusiness in North and South America. Other than that, Europe, Asia, Africa and Australasia all said no.

Today the situation is one of global stalemate. GMOs are not going away, but any moves to introduce new crops or move into new areas are bitterly contested. To me it resembles the First World War in about 1916. The first climactic battles have already happened, and no one has yet won. New battlefronts come and go, but the trenches barely move from one year to the next. In what has become a lengthy war of attrition, both sides have resorted to underhand tactics and have demonised their opponents in propaganda. Sooner or later there will have to be peace negotiations and an Armistice – but how many more years must we lose in the meantime? The first step must surely be understanding your enemies, and to the fullest extent possible recognising the justice of their cause.

CHAPTER EIGHT

What Anti-GMO Activists Got Right

The process of writing this book has been unexpectedly cathartic. When I began this project I intended it as an exposé of sorts, a passionate polemic highlighting the sheer injustice and irrationality of the movement against genetic engineering and the damage I believed it had done in the world. Accordingly I planned, and even fully wrote, chapters exposing the funding sources of anti-GMO groups, holding their leaders up to ridicule, debunking the pseudoscientific basis of most of their claims, and demonstrating conclusively how genetic engineering can both be of service to the environment and improve the lives of farmers in poorer countries.

I had this book all ready to go, and even sent a draft to my publisher. But something in me also held back. I was aware that, entertaining as it might be, and reassuring to anyone who already agreed with my self-professed pro-science Damascene conversion, I wasn't really doing much service to the cause of truth – my analysis was shallow, many of my targets were straw men, and if I were advancing any cause at all it would be one of polarisation rather than illumination. To return to my First World War analogy, I was like a deluded general, putting my all into moving the trenches another mile into no man's land, and damn the casualties. No thought of a ceasefire, let alone a negotiation or any attempt at genuine understanding of the other side, entered my mind. Such is the mentality of perpetual warfare. And it was even more peculiar in my case as I had once been among the very people, who – glimpsed across the mud and through the drifting smoke of the battlefield – I now saw as 'the enemy'.

Some of these types of angry, one-sided books have already been written, of course. If you really want to try one, I recommend Henry Miller's *The Frankenfood Myth: How Protest and Politics Threaten the Biotech Revolution*. You don't really need to read past the title; but if you do, you will come across lots of descriptions of activist groups as 'irrational', 'anti-technology', trying to hold up progress because they simply don't understand scientific facts. My first draft wasn't quite this extreme, not least because I don't share Miller's pro-business politics. I also still had split loyalties. I couldn't in good conscience dismiss people as 'irrational' and thoughtlessly 'anti-technology' when they were still my friends and I knew them as smart, rational, moral human beings. Accordingly, I sent the book draft out to some of the people I knew would be most critical of it, and asked for comments. I also plucked up the courage to ask for interviews with those I had become pitted against in recent years. So although I first wrote the book that way, at some deeper level I didn't want to end up publishing narrow heroes-and-villains caricatures of reality, merely the flipside of what you might read in anti-GMO treatises about the evilness of Monsanto. This was not a time-efficient process: I deleted entire chapters, watching tens of thousands of carefully crafted and meticulously researched words representing months of work disappear with the push of a button.

What my friends helped me realise was missing from my first draft was an honest appreciation of where the anti-GMO movement came from, what it really stands for, and what drives the people in it who have genuine concerns about genetic engineering. You would think that, as a former activist against genetic engineering myself, I would have some insight into this. But the vagaries of memory, oiled by the mind-easing bias of retrospective justification, are such that I had long since 'deleted' the feelings and world view that I used to have when I was busy slashing down field trials back in the 1990s. Moreover, there were times when I really hated, and was in turn detested by, my former friends on the anti-GMO scene. I pretended not to be, but I was wounded when people I had once loved and respected began calling me a liar. Some even asked me, in all seriousness, to prove that I wasn't receiving corporate cash from the biotech industry. Jim Thomas, through whom I had originally been introduced to the whole subject back in 1996, told the *Observer* in March 2013 that I had 'built a very successful career on the back of portraying people who were his friends as unthinking'.[1] Whether that's true or not, I didn't want this same accusation to be levelled at this book with any justice.

Some occasions are almost too painful to recall. The absolute worst moment of all was probably the time Stewart Brand and I, following the screening of a Channel 4 documentary featuring us both entitled *What the Greens Got Wrong*, had to face a :dio debate against my Oxford friends George Monbiot and Doug Parr from Greenpeace. Given the content of the programme, which accused the environmental movement of having harmed humanity with its opposition to genetic engineering, nuclear power and the use of DDT to control malaria, it is perhaps not surprising that the subsequent discussion was ill-tempered and unpleasant. That was the point during my entire experience that I came closest to any actual physical threat: one member of the audience literally squared up to me and shouted in my face outside the studio during a break. There was another hostile crowd watching on screens outside the studio; I almost ran past them, as if I were a guilty convict leaving court with a coat over my head, desperate to get away from a baying mob. Stewart and I had dinner later with a friend of his; I was barely able to hold up my end of the conversation as I kept glancing around, unable to concentrate for a moment, since we were only a short distance away from the venue in London. I could barely sleep that night or the next. I remember feeling both embattled and alone.

George is not one to mince words, and he wrote about the Channel 4 programme in the *Guardian*: 'Brand and Lynas present themselves as heretics. But their convenient fictions chime with the thinking of the new establishment:

corporations, thinktanks, neoliberal politicians.'[2] Needless to say, having been friends for nearly 15 years by this point, we found we no longer had much to say to each other. I began to see George as a one-dimensional character, a hypocrite who was simply stuck in a knee-jerk tribal defence of the green movement he was so emotionally attached to. He, as the sentence above suggests, had begun to see me as a tool of big business, wittingly or unwittingly part of what he described in the same article as a 'powerful counter-movement' to environmentalism, an effort 'led by corporate-funded thinktanks' which had long 'waged war on green policies'. George had spent his life fighting these kinds of people. And now it appeared to him that I was one of them, or at least fronting one of their campaigns. We had furious rows over email and, on one particularly memorable occasion, over what was supposed to have been a score-settling lunch. To his credit, however, George both stuck to his guns politically and refused to let our friendship die. Our lines of communication were both narrowed and strained for several years, but not closed completely.

George later told me why he felt the need to take the matter so seriously, because Stewart and Channel 4 were repeating what he considered a 'blood libel' against the environmental movement. This was that its opposition to the use of DDT in agriculture had led to a de facto ban on spraying in developing countries against the mosquitoes that carry malaria. Although at my behest the most overblown allegations had been cut from the programme before broadcast, Stewart's book *Whole Earth Discipline* contained a quote from a US public health official claiming that 'the ban on DDT may have killed 20 million children'.[3] I have never seen any convincing evidence to support this claim, which has indeed, as George asserted, been eagerly promoted for years by anti-environmental campaigners promoting climate change denial and the interests of corporate lobbyists. It's a shame the programme repeated this myth: not just because it overshadowed serious debates we could and should have had about nuclear power and GMOs, but also because it added such a poisonous dimension to my relationships with people I cared about.

The most serious rupture was with the man who for many years had been my soulmate, Paul Kingsnorth. I had known Paul, a novelist, poet and former deputy editor of the *Ecologist* magazine, since the late 1990s. We had been best man at each other's weddings, and had supported each other through thick and thin during long nights in Oxford's pubs sipping real ale and grumbling about the state of the world. Our falling out was about much more than merely GMOs. While Paul had moved more towards deep ecology and set up an artistic project of 'uncivilisation' called Dark Mountain, I had become more optimistic, technocratic and an 'eco-pragmatist', as I presumed to call it. In his *Dark Mountain Manifesto*, Paul (and his co-author Dougald Hine) wrote: 'We tried ruling the world; we tried acting as God's steward, then we tried ushering in the human revolution, the age of reason and isolation. We failed in all of it, and our failure destroyed more than we were even aware of. The time for civilisation is past. Uncivilisation, which knows its flaws because it has participated in them; which sees unflinchingly and bites down hard as it records – this is the project we must embark on now.'[4] The mention of 'God's steward' was, I assumed, an allusion to Stewart Brand's famous quote: 'We are as gods, and might as well get good at it,' which I took as the inspiration for the title of my 2011 book *The God Species*.

It is hardly surprising that Paul and I fell out, since our philosophical journeys took us over many years in such starkly different directions. As Paul was rejecting rigid rationality and heading out into the wilds to re-immerse himself in nature, I was on the West Coast of the US, sitting down with Stewart Brand and the proudly eco-heretical Breakthrough Institute to draft the polar opposite viewpoint, our *Ecomodernist Manifesto*. This declared, somewhat grandly, 'We call ourselves ecopragmatists and ecomodernists. We offer this statement to affirm and to clarify our views and to describe our vision for putting humankind's extraordinary powers in the service of creating a good Anthropocene' (the Anthropocene is the name given to our new human-dominated geological epoch). While we were all ostensibly aiming for the same environmental ends, Paul argued for humility, while our ecomodernist approach seemed to him like hubris: we wanted, as our *Manifesto* declared, 'humans [to] use their growing social, economic, and technological powers to make life better for people, stabilize the climate, and protect the natural world'. Paul's narrative was one of retreat and acceptance, which I found baffling or even borderline nihilistic. To him, our cold statistics and our god-posturing made us part of the problem, not the solution.

I don't have to try to explain Paul's world view and his attitude to us ecomodernists because he did so himself in various eloquent critiques and essays.* Writing in the *Guardian* in 2012, Paul described how he saw our newly minted philosophy.

> *Neo-environmentalism is a progressive, business-friendly, post-modern take on the environmental dilemma. It dismisses traditional green thinking, with its emphasis on limits and transforming societal values, as naive. New technologies, global capitalism and Western-style development are not the problem but the solution. The future lies in enthusiastically embracing biotechnology, synthetic biology, nuclear power, nanotechnology, geoengineering and anything else new and complex that annoys Greenpeace.*

But, he conceded, 'Neo-environmentalism is beginning to make waves in certain circles. Stewart Brand gives talks all over the world arguing the case for megacities and GM crops; British writer Mark Lynas gets airtime to promote nuclear power and attack his former green friends as "Luddites".' Paul

---

* Whatever our differences I could never deny that Paul is an excellent writer; his novel *The Wake* was deservedly longlisted for the prestigious Booker Prize.

was critical of conventional environmentalism too, because 'its language and its focus have grown increasingly technocratic and scientistic ... Any campaign to protect the wild world which avoids acknowledging our intuitive, emotional relationship with it will leave itself open to the kind of heartless ideological assault it is now receiving from the neo-greens.'

I was somewhat annoyed about this description of our endeavours, not least because the term 'neo' seemed in itself to be pejorative – think neoliberals, neocons, or worse still, neo-Nazis. But the problem I had was that in hating Paul – and we really did loathe and detest each other, for the best part of a decade – deep down I knew that I was also hating something in myself, something that I used to be and might one day like to be again. While ecomodernism made full intellectual sense to me as we sat around a conference table in San Francisco, I didn't want to completely cut myself off from my past. My head might be with the ecomodernists, but my heart was still with Paul, somewhere out in the woods and mountains far away from spreadsheets and scientific papers.

I still think it was wrong of critics like Paul and George Monbiot to see our manifesto as anti-environmental: we were obsessed with protecting wilderness, but our vision was one of 'decoupling' human dependence on nature in order to allow ecosystems to recover. Cities are therefore good, because they are a more efficient way to run large-scale human habitation than spread-out low-intensity developments; the manifesto website was illustrated with a photograph showing a background of towering city blocks behind a foreground of green forest (I think it was taken in Hong Kong). The contrast was stark, but it suggested separation, alienation even, between humans and nature. Our manifesto was empirically well-grounded but, for many, emotionally off-putting. Not surprisingly most greens saw reflected in our vision a chilly dystopian future of skyscrapers and lab-meat, where our planet-bestriding technology might protect the wilderness that remained behind the glass but would in the process disconnect us from something deeper about what it means to be human on this Earth.

One of the criticisms that environmentalism has long made of modern society is that it severs people's connection with nature. This is something that has always bothered me too. There's no comparison, for example, between swimming in a pool and swimming free in the river. I live close to the Thames, and often bike down to a bend in the river where it passes the western side of Port Meadow, a centuries-old area of common land grazed by cows and horses. While in a swimming pool you are confined by straight lines – the squares of the tiles, the rectangle of the pool itself, swimming backwards and forwards as if in a liquid treadmill – in the river you have true freedom of both movement and of mind.

There's also much more to see. On one recent swim the wind was scudding across the water, stirring up small ripples as I waded through the mud and weed to the deeper central channel of the river. A sharp-eyed Arctic tern hovered briefly above the shallows and then dived with a tiny splash to catch a fish. Iridescent blue damselflies skittered above the surface, while a lone jet-black rabbit grazed under hawthorn trees on the far bank. I could see mint growing down to the water line, while little yellow knots of lilies gathered on the surface. To the various animals, humans seem perhaps less threatening when submerged and at eye level; swans, which normally raise their wings and hiss if I get too close, drifted along unconcernedly a few feet away. A white egret stalked about on the shore. One of my dogs, loyal as always, paddled along beside me, breathing heavily through its nose. The other dashed about in the undergrowth. Just before I returned to the bank, a lone wasp inadvisedly dropped to the water's surface. There was a plop as a fish rose unseen from the depths, and it had gone.

A swimming pool is controlled, clean and chlorinated, like the human-managed Earth our ecomodernist vision seemed to present to critics. Even if a tangled jungle flourishes beyond the fence, it is not a nature-immersing experience. Swimming up and down between the roped-off lanes, there is no experience outside the rigid boundaries of modern life, no appreciation of oneself as just another animal sharing this beautiful planet. Every time I go for a swim in the river on the other hand, or a run in the Black Mountains – which I pine for every time I spend too long stuck in the flatlands of Oxford – I am reminded that I don't want to disconnect myself from nature. Like Paul I don't want all my life to be mediated by modernity or technology. I'm still much more eco than I am modernist. It's the very timelessness of the river that makes it an experience of transcendence. I return with muddy feet each time, but spiritually refreshed.

In 2011 Paul wrote an essay for *Dark Mountain* that probably comes closest to describing the deeper reason for our rupture. It is called 'The Quants and the Poets'. I am generally a quant (that is, a quantifier). Paul is generally a poet. This is a literal as well as a metaphorical description of our different personalities. Paul actually *is* a poet, and has several published volumes of poetry to his name to prove it. I have not written a poem since I was forced to express myself in this unnatural way at primary school. Even my teenage heartbreaks were never expressed poetically, thank goodness, unless you count hours alone in my room singing along to The Smiths. These days I am much happier perusing scientific papers. Although I'm not much good at maths, I get a kick out of numbers. I don't feel I can understand a problem fully until it is properly quantified. I love to understand the world empirically, from the precise heights of mountains to the names and altitudes of clouds and the ages of rocks. For me this doesn't detract from the awe and majesty of nature. If I know the striations on a rock outcrop near Snowdon in North Wales were carved by glaciers during the Last Glacial Maximum 200 centuries ago, and that the billowing cloud growing near the 1,085-metre summit is a cumulonimbus, then I'm all the happier that the landscape and cloudscape I am appreciating carries this extra dimension of meaning thanks to science. These differences of approach – the cause of much humorous teasing on both sides – initially deepened our friendship, rather than divided us.

Paul's essay, written just after the accident at Fukushima in Japan, was ostensibly about nuclear power, but in reality was about something deeper. 'My feeling is that the green movement has torpedoed itself with numbers. Its single-minded obsession with climate change, and its insistence on seeing this as an engineering challenge which must be overcome with technological solutions guided by the neutral gaze of Science, has forced it into a ghetto from which it may never escape. Most greens in the mainstream now spend their time arguing about whether they prefer windfarms to wave machines or nuclear power to carbon sequestration. They offer up remarkably confident predictions of what will happen if we do or don't do this or that, all based on mind-numbing numbers cherry-picked from this or that "study" as if the world were a giant spreadsheet which only needs to be balanced correctly.'

That – minus the cherry-picking, thank you, Paul – was a pretty accurate description of what I had been up to for the past several years, in trying to figure out how much different technologies like nuclear power and renewables could contribute to tackling climate change. In some work I did with the late scientist, David MacKay (to whom this book is dedicated), we literally did turn the world into a giant spreadsheet – a gigantic Excel document where hundreds of columns of numbers represented the climate impact of our combined choices in the global economy. The idea was to present a 'global calculator', where different options – How much nuclear? What area of windfarms? What crop yields? – could be quantified on a global scale and used to develop and present a pathway in the future that had at least some anchoring in empirical reality. The original inspiration for this came in David's classic book *Sustainable Energy Without the Hot Air*, where, tiring of endless ideological battles about energy technologies, he had presented them as quantified choices. One of his classic lines, which I never tired of quoting, was 'I'm not pro-nuclear, I'm pro-arithmetic'.* His overall point was that it was fantasy to imagine that the world could be easily decarbonised with the simple addition of a few solar panels and windmills. Without numbers our endless arguments could never get anywhere.

Paul argued that this sort of effort was missing the point. His essay, which really deserves reading in its entirety, went on to say:

> The fight between the pro-nukers and the anti-nukers, for example, is actually quite archetypal. Though both sides pretend to be informed by 'science' and 'facts' both are actually informed primarily by prejudice. Whether you like nuclear power or not is a reflection of the kind of worldview you have: whether you are a confident embracer of the Western model of progress or whether it frightens or concerns you; whether you trust science or tend not to; whether you are cautious or reckless; whether you are 'progressive' or 'conservative'. On issues ranging from GM crops to capitalism, these are the underlying stories that actually inform the green debate … A green quant might be telling you to change your lightbulbs or come out on the streets in favour of a nuclear power plant or a windfarm, but he's not asking you to examine your values or your society's underlying mythology.[5]

Paul was careful to acknowledge that no one is fully quant or fully poet, and that actually this is 'a tension that is present within all. None of us is wholly, or even primarily, rational and analytical, and none of us is quite devoid of poetry either, though it is sometimes hard to find it.' He concluded: 'We have no shortage of arguments about numbers and machines,

---

* David MacKay was also a Cambridge University mathematician who wrote a textbook on Bayesian statistics, which he dedicated to the Campaign Against the Arms Trade.

but we do have a great shortage of workable stories.' It was time, in other words, for the quants to step back and let the poets take the lead.

Finding the right balance between the emotional and the rational is something that has troubled deep thinkers like Paul for a long time, and was at the centre of debates about the proper role of human reason during the Enlightenment. In deference to the poets, rather than trying to explain it in prose, I will switch instead to verse (which I do, by the way, sometimes appreciate).

> *Do not all charms fly*
> *At the mere touch of cold philosophy?*
> *There was an awful rainbow once in heaven:*
> *We know her woof, her texture; she is given*
> *In the dull catalogue of common things.*
> *Philosophy will clip an Angel's wings,*
> *Conquer all mysteries by rule and line,*
> *Empty the haunted air, and gnomèd mine—*
> *Unweave a rainbow, as it erewhile made*
> *The tender-person'd Lamia melt into a shade.*

'Philosophy will clip an Angel's wings.' I could have spent 3,000 words trying to explain that and not even come close. John Keats did it in six. His poem *Lamia*, published in 1820, is perhaps the best exposition of the tension between Romanticism and Reason that Paul was also exploring a variant of in his essay.[*]

So does 'cold philosophy' (philosophy in Keats' time meant 'science' rather than the broader meaning we have today) really unweave the rainbow? The question has troubled scientists too. Richard Dawkins, in a book titled *Unweaving the Rainbow*, after Keats' poem, declared that

---

[*] To call someone 'romantic' is now used as a term of abuse, as if to illustrate the deadness of our culture that Paul was drawing attention to.

'Keats could not have been more wrong' in believing that Newton had 'destroyed all the poetry of the rainbow by reducing it to prismatic colours'. Isaac Newton, using a prism, was the first to discover that visible light could be split into the colours of the rainbow as part of the wider electromagnetic spectrum. Dawkins writes, 'Newton's unweaving of the rainbow led on to spectroscopy, which has proved the key to much of what we know today about the cosmos. And the heart of any poet worthy of the title Romantic could not fail to leap up if he beheld the universe of Einstein, Hubble and Hawking.'

While Dawkins appreciates good poetry too, he worries that the sense of wonder and awe that all of us feel can be easily perverted, in the hands of lesser people than poets, in the direction of cheap mysticism and superstition if not combined with sufficient scientific education. He therefore tries to resolve the tension between the rational and the romantic by insisting that they can be one and the same if directed properly. Dawkins quotes William Blake's famous opening quatrain in *Auguries of Innocence*:

> *To see a World in a Grain of Sand*
> *And a Heaven in a Wild Flower*
> *Hold Infinity in the palm of your hand*
> *And Eternity in an hour...*

Dawkins then writes: 'The impulses to awe, reverence and wonder which led Blake to mysticism (and lesser figures to paranormal superstition) are precisely those that lead others of us to science. Our interpretation is different but what excites us is the same. The mystic is content to bask in the wonder and revel in a mystery that we were not "meant" to understand. The scientist feels the same wonder but is restless, not content; recognizes the mystery as profound, then adds, "But we're working on it."'[6]

Dawkins is contemptuous of the academic fad of post-modernism. He quotes the American anthropologist Matt Cartmill to sum up the basic idea: 'Anybody who claims to have objective knowledge about anything is trying to control and dominate the rest of us ... There are no objective facts. All supposed "facts" are contaminated with theories, and all theories are infested with moral and political doctrines ... Therefore, when some guy in a lab coat tells you that such and such is an objective fact ... he must have a political agenda up his starched white sleeve.'[7] As the French plant scientist Marcel Kuntz has complained: 'The central goal of science, defining what is true and what is false, becomes meaningless they [the postmodernists] argue, as its objectivity is reduced to "claims" that are simply the expression of one culture – one community – among many. Thus, all systems of thought are different "constructs" of reality and all additionally have political connotations and agendas.'[8]

Some environmentalists have come close to embracing postmodernism via their suspicion of what they see as the inherent biases and power relations embedded in modern science. One such is Vandana Shiva, the Indian writer and campaigner I mentioned in a previous chapter, who wrote in one of her books that Western science is merely 'a local tradition that has been spread worldwide through intellectual colonisation'. To Shiva, 'indigenous traditions of knowledge' tend to collapse in the face of scientific thought not because they fail to explain the universe as effectively as Newton, Darwin and Einstein, but because they are violently erased by colonialism as if at the point of a gun. She may have a point here: more politically assertive religions like Christianity and Islam seem to be doing very well despite the simultaneous existence of modern science. To Shiva the classical scientific method pioneered during the European Enlightenment is therefore not so much a system of knowledge generation as of power propagation. Not only does it have no special claim to objective truth, but all scientific claims to universal truths are suspect by definition, for they are in reality the instruments of neo-colonial power.

Shiva might be taking this too far, but she isn't entirely wrong. Knowledge is indeed power, insofar as understanding how something works in nature pretty much by definition gives humans the power to amend and alter it. Understanding the nuclear forces in the heart of the atom was undeniably the first step along the path towards nuclear fission, whether in the bomb or the reactor. Genetic engineering is surely another example. When Francis Crick and James Watson opened up the science of molecular biology by unravelling the structure of DNA, they began a process which – perhaps inevitably – ended up with the shuttling of genes between species for human purposes, and more latterly precise gene editing. As documents from the time show, all the pioneers of genetics understood this perfectly well. Once it had become clear that DNA was the same coding system in all living organisms, then it was equally clear that different sequences called 'genes' should function similarly wherever they happened to be, whether in a rabbit or a bacterium. While many of these pioneers were concerned initially that recombinant DNA – especially where genetic material had been transferred across the species barrier – might harbour some unanticipated ill effects, the fears of most subsided when further research showed this was very unlikely to be the case.

While I agree with Kuntz that the postmodernist rejection of objectivity is itself objectionable, I cannot completely reject the idea that scientific work can have, as he puts it, 'political connotations and agendas'. It is indisputable surely that science and technology are two sides of the same coin, and that one implies the other. If knowledge is power, by the nature of our unequal society some people will be more empowered by knowledge and its resulting technological applications than others. This was a point made forcefully to me by George Monbiot when I finally sat down and interviewed him formally under the apple tree in his Oxford garden in July 2017. We were discussing whether or not science was values-free. 'Nothing human beings do is values-free!' George insisted between munching the ripe cherries that I had brought in lieu of payment. 'We are saturated in values. And the decisions we make are always going to be driven by our social environment, by our political environment, by the values which we hold without necessarily even being aware of, and those decisions include the decisions as to what we study and how we study it.' He spat out a cherry stone. 'They don't alter the scientific methodology that you might follow, but they alter the question of what you apply the scientific methodology to. We cannot leave science to the scientists alone,' he concluded. 'Cherry?'

I suggested that this put him in the same relativist camp as the postmodernists. George didn't think so (he is, incidentally, more of a real scientist than me, having studied zoology at university). 'I'm very much in favour of the scientific method,' he told me, 'but I'm in favour of ensuring that the questions that the scientific method is used to determine are informed by as broad a worldview as possible. And scientists, the *unusual* ones have a broad worldview, most of them that I meet tend to have very narrow perspectives, and that's a real problem when you're framing your fields of enquiry.' But didn't this, I persisted, imply a rejection of the whole scientific method? 'No, what the scientific method seeks to do is to impose a discipline which can be followed, and should be followed when you are trying to determine the answer to a scientific question. Of course the scientific method could be potentially described as value-free, even though that is contestable, but science itself is saturated in values – Enlightenment values, economic values, a whole series of values which scientists often fail to identify to themselves.'

So to accuse opponents of GMOs as being generically 'anti-science' makes no sense at all to George. 'I think that was a ridiculous designation … it's a bit like saying that people who are anti-chemical weapons are anti-chemistry, or who are opposed to nuclear annihilation are anti-physics. GM, like washing machines or cars, is a technology, and we have to make a political decision – which in an ideal world would be a democratic decision – as to *whether* we want to use it or not and the *extent* to which we want to use it or not, and *how* we want to use it or not. And to dismiss all that as being anti-science was absurd, facile and stupid. But it was a meme cultivated very often by the corporations themselves because it was an easy way of dismissing their opponents as irrational.' I coughed self-consciously, thinking back to all the many blog posts and tweets I had written accusing my opponents of being 'anti-science'.

So is anyone actually 'anti-science'? Interestingly George did feel that the term might be applicable to climate scepticism in some cases. 'Yes, I think there are certain elements of climate change denial which are anti-science. Where they're basically attacking the scientific establishment and the very notion of peer review. I mean that seems to me to be anti-science. That's not what we [early genetic engineering opponents] were doing. And they're not attacking a technology, they're attacking the findings. They're saying climate change isn't happening, and all the people who say climate change is happening are scientific frauds who have deliberately fiddled their results.' And so when it came to the science of genetic engineering, 'there was definitely a problem, I felt, with the direction that research was taking, but it wasn't that the research itself was wrong, or that anyone was faking the results, it was just that I felt they were asking the wrong questions.'

The issue of political power, George emphasised, lay at the heart of his concerns about the development of genetically modified crops, both back in the 1990s when he spoke and wrote so forcefully against them, and today. The inherent food safety issue of GMOs was 'never a concern of mine', he told me. Nor was the idea, put about by Prince Charles in what George described scornfully as a 'deeply misguided, poorly argued, confused speech', that humans risked playing God by moving genes around. He was very happy to concede, moreover, that 'it is absolutely true that there's a scientific consensus on GMO safety'. But even though this was central to my own change of heart on GMOs, it was not at all the key point for George. 'For me it was all about corporate power, patenting, control, scale and dispossession.' These issues were particularly concerning at the time because of Roundup Ready, which he viewed as 'scorched earth farming, where you can develop an entire farm of soybeans with no other plants growing at all because Roundup will take out everything else', and the resulting dominance of Monsanto in the deployment of genetic engineering, plus the legal abolition of farmer control over seed-saving through commercial patents on genes.

This reminded me of a comment by the American environmental writer Nathanael Johnson in the movie *Food Evolution*, that genetic engineering 'is just a technology … It doesn't have any moral valence.' I put it to George that technologies could be used in different ways by different people, and even if exploited by a private corporation in one place could still be used in another to advance a public good. Genetic engineering examples might be drought-tolerant maize or disease-resistant bananas developed by public sector projects in Africa. Looking at the technology today, George admitted that he was quite prepared to 'take into account the fact that it's not the same thing as it was when I started campaigning against it … And I can see that there are some publicly funded projects that could have public benefits.' But, he went on, 'I would still ask some searching questions of it, like how does this change the balance of power between small producers and large producers, between rich and poor, between rich nations and poor nations, between corporations and people, between public agencies and people. The balance of power is a crucial issue which is all too often neglected when it comes to public policy, because public policy believes that it's driven by a rational assessment of technology, and of economic progress as it calls it, and of economic growth, but all these things are saturated with power relations. And if you pursue issues like that when you're blind to power relations you're only going to reinforce dominant power.'

Researching this issue further, I came across a good illustration in support of George's argument from Paraguay, in South America. According to the development agency Oxfam, land distribution in Paraguay is the most inequitable in all of Latin America, with 80 per cent of agricultural land being held by just 1.6 per cent of landowners. That leaves 300,000 family farmers with no land at all, contributing to one of the worst poverty rates in South America.[9] Landless farmers who have occupied or agitated for land reform have been imprisoned or even murdered – 129 have been extra-judicially executed since the end of Paraguay's long-running dictatorship in 1989. The majority of Paraguay's agricultural land is now dedicated to soy production, mostly Roundup Ready. Oxfam presents evidence demonstrating how the Roundup Ready system privileges larger farmers over smaller ones, thereby intensifying inequalities of land.

Oxfam writes: 'Large-scale monoculture expansion is competing for land with small-scale basic food production; thus, households which used to be self-sufficient in food now rely on local markets, where nutritious food is not always available.'[10] The Roundup Ready system therefore acts in Paraguay as an economy of scale, helping the big producers to out-compete the small – and perhaps even displace them altogether. 'Large-scale monoculture expansion, driven by world market dynamics and financial interests, tends to deepen the concentration of land ownership, limit equitable access to resources, degrade the environment, harm the health of the local population, create exploitative working conditions and put at risk the traditional livelihoods of small-scale farmers,' Oxfam's report concludes. There are surely many other examples around the world, and while George's political economy objection is not a fundamental argument against all genetic engineering, I was happy to concede under his Oxford apple tree that he was making a convincing case. Plus, the cherries had run out, and it was time to go inside and cook lunch.

'Have you ever read Langdon Winner's book, *The Whale and the Reactor*?' asked Jim Thomas, the first time we spoke in over a decade. I confessed I hadn't. 'To me that's the classic argument against nuclear power, where he basically says the nuclear reactor can only exist in a centralised state with large amounts of centralised power because you need to defend it, you need security infrastructure and all this stuff; the technology cannot be a democratised technology, it requires centralised systems. And I sort of feel that's true for biotech too, the technology's constitution lends itself to being best exploited by well-capitalised institutions.'

It was a strange feeling to be talking again to the man who, having first influenced me into campaigning against genetic engineering back in the mid-1990s, had almost become my nemesis after I changed my mind. Although I began by admitting to having felt a little nervous before our Skype call (Jim lives in Quebec with his wife and two children), soon we were talking animatedly, almost – as we should perhaps have been – like old friends. Jim Thomas was someone I particularly wanted to talk to, not just because he had a big role in my personal journey, but because he has had a consistent and impressive career as a social justice campaigner and technology critic. After starting out at Greenpeace soon after leaving university (where like me he studied history), about ten years ago he moved to the ETC Group, which, as he put it, 'is this weird acronym, Erosion, Technology and Concentration'. According to him ETC Group's work focuses 'around how the combination of emerging technologies with corporate concentration can lead to various erosion processes: biodiversity erosion, erosion on human rights, erosion on democracy, that's the sort of fundamental question that we are exploring. We are trying to understand how patterns of corporate concentration combine with emerging technologies to change power, basically.'

It's not all about GMOs by any means. When I spoke to him Jim was working on blockchains, the internet database system behind crypto-currencies such as bitcoin. Other recent work has focused on geoengineering (intentional climate manipulation) and synthetic biology. On agriculture, ETC Group has a forward focus and broad analysis about overall trends that makes its work unusual among campaigning groups. One recent ETC briefing paper, entitled 'Software vs. Hardware vs. Nowhere'[11] examined corporate mega-mergers, which it said 'will transform the entire world agricultural inputs industry – from crop and livestock genomics to farm machinery and insurance'. If the deals go through, an ETC spokesperson was quoted as saying, 'it is inevitable that the farm machinery companies will step in to merge their Big Data machinery with Big Data genomics'. This was all a far cry from the simplistic anti-Monsanto activism I had imagined most anti-GMO campaigners were still engaged in. Moreover, as he talked me through his concerns, I began to better understand Jim's point on a number of issues.

I also followed his advice and read *The Whale and the Reactor*, a book first published in 1986 whose subtitle is *A Search for Limits in an Age of High Technology*. 'If the experience of modern society shows us anything,' Winner suggests early on in the book, 'it is that technologies are not merely aids to human activity, but also powerful forces acting to reshape that activity and its meaning.' Technologies may therefore not have an inherent 'moral valence', in Nathanael Johnson's phrase, but they can certainly have important social and political implications. 'Individual habits, perceptions, concepts of self, ideas of space and time, social relationships, and moral and political boundaries have all been powerfully restructured in the course of modern technological development,' Winner writes.[12]

Of course, to argue that a tool like a hammer is inherently 'political' seems ridiculous, unless perhaps you are a nail. As Winner explains: 'To discover either virtues or evils in aggregates of steel, plastic, transistors, integrated circuits, chemicals, and the like seems just plain wrong, a way of mystifying human artifice and of avoiding the true sources, the human sources of freedom and oppression, justice and injustice. Blaming the hardware appears even more foolish than blaming the victims when it comes to judging conditions of public life.' But although he resists simplistic 'technological determinism' – the notion that certain tools necessarily lead to certain related social or political outcomes – Winner does conclude that 'artefacts can contain political properties'. He ends the book with a paean against nuclear power as perhaps, like Jim Thomas also told me, the classic example of a technology with inherent political characteristics – its need for centralised control and safeguarding, its legacy of radioactive waste and its production of isotopes that can also be diverted to weapons production.

'Yes, we may be able to manage some of the "risks" to public health and safety that nuclear power brings. But as society adapts to the more dangerous and apparently indelible features of nuclear power, what will be the long-range toll in human freedom?' Winner asks. He quotes environmentalist Denis Hayes, the founder in 1970 of Earth Day, who once claimed: 'The increased deployment of nuclear power facilities must lead society toward authoritarianism. Indeed, safe reliance upon nuclear power as the principal source of energy may be possible only in a totalitarian state.' Hayes was later proven wrong, in point of fact: during the 1980s France converted to almost entirely nuclear electricity without abandoning its democratic character in the process. But on the other hand I can see his point: unlike so-called 'soft energies' such as solar and wind, nuclear fission does not lend itself to small-scale, low-tech applications. Nuclear reactors, pretty much by definition, are not built by village cooperatives. In fairness, solar photovoltaic panels aren't manufactured in rustic wooden-beamed village cooperatives either – they are produced on an epic scale in large factories, even though they can then be installed in a much smaller setting.

Jerry Mander, the one-time advertising executive who became a potent critic of high technologies and globalisation during the 1970s, made pretty much the same case against nuclear power, which he confidently declared 'cannot possibly move society in a democratic direction, but *will* move society in an autocratic direction. Because it is so expensive and so dangerous, nuclear energy must be under the direct control of centralized financial, governmental, and military institutions. A nuclear power plant is not something that a few neighbors can get together and build. Community control is anathema.'[13] Like ETC Group and Langdon Winner, Mander was consistent enough not to confine himself to criticising one single technology. Indeed his first major work was the catchily titled 1978 book *Four Arguments for the Elimination of Television* (you can tell he was an ad-man). The book made clear that this title was intended to be taken entirely literally. As with nuclear power, Mander claimed that the inherent political nature of television as a medium would lead to the erosion or even extinction of democracy. This was because it was now possible, he wrote, for one powerful person or institution to speak to 'an entire nation of 200 million people' as individuals via the television set. He insisted that the conditions of television viewing, 'confusion, unification, isolation' combined with 'passivity', were 'ideal preconditions for the imposition of autocracy'.[14]

Mander used the analogy of the motorcar to illustrate his case. If you accept the existence of automobiles, he wrote, this also implies road infrastructure, an oil industry, as well as 'a sped-up style of life and the movement of humans through the terrain at speeds that make it impossible to pay attention to whatever is growing there'. In the same way, 'television itself predetermines who shall use it, how they will use it, what effects it will have on individual lives', and – chillingly – 'what sorts of political forms will inevitably emerge'. There could be no compromise with the existence of television, Mander concluded. Either it must be 'gotten rid of totally' or democracy itself would wither.

It is stimulating to read this polemic almost 40 years later, when television has ceased to be the mass experience of communal passivity that Mander so worried about in 1978. Today, hardly anyone I know sits down and watches TV, and certainly not together. Electronic screens are just as dominant in our lives as before – perhaps even more so – but with the coming of the internet, mass communication seems to have been more affected by centrifugal forces rather than centralising ones. This adds another dimension to the whole debate, the dimension of time, as technologies morph from one thing to another in a process of constant reinvention and change. What television meant in 1978 it surely no longer means in 2017. Mander wrote: 'To speak of television as "neutral" and therefore subject to change is as absurd as speaking of the reform of a technology such as guns.' But television did change – and it did not lead to the automatic 'imposition of autocracy'. Although it was certainly used as a tool by autocrats the world over at various times, at other times it helped draw attention to injustice and advanced the cause of freedom. Even guns, a technology of death, can be used either to maintain a police state or to remove a hated dictator.

This all begs the question of whether one can ever tell in advance how a technology will turn out. There are numerous examples of how the search for one tool has led to the subsequent evolution or invention of another, often with quite different implications. As Tim Harford writes in his book *Fifty Things That Made the Modern Economy*, radar was first developed out of the British military's search (during the run-up to the Second World War) for a 'death ray' utilising focused radio waves. The ubiquitous modern device, the iPhone – on which, incidentally, I recorded my interviews with both Jim Thomas and George Monbiot – depends on at least 12 separate contributor technologies which were first developed in the public sector for entirely different purposes, mostly with support from the US government, and often from the Department of Defense. These include GPS, or global positioning system, as Harford writes: 'a pure military technology, developed during the Cold War and opened up to civilian use only in the 1980s'; the web, developed by software engineer Tim Berners-Lee while he worked in particle physics research at CERN; and of course the internet itself, which famously began as ARPANET, a network of computers funded by the Pentagon and originally conceived as a distributed command-and-control network that might survive a nuclear war. The same goes for other components such as touch screens, algorithms, HTTP and Siri.

Given all this experience, I asked Jim Thomas on our Skype call – using a technology that would have seemed like a Star Trek fantasy at the time Jerry Mander was writing his lament about television – whether we can really conclude anything about the inherent political natures of different technologies? Hasn't the mobile phone, after all, been taken up as a liberating tool by just about every farmer in sub-Saharan Africa? 'I actually do have concerns about the massive rollout of internet infrastructure across Africa and buying everyone a cellphone and running energy systems and agriculture systems and what power that gives to initially telecoms companies but later big data companies,' Jim replied. 'It's not always useful to see things from the short-term point of view of an individual, and the promises they've been told; it's often better to zoom out and ask how does this change agriculture, how does this change farming systems over a period of time and then where does the individual find themselves?'

As to biotechnology itself, Jim was happy to concede that there is a real debate to be had about democratic, public sector – even open-source – applications of genetic engineering. 'I think it's an interesting question which side of the line it falls on. Is it ultimately going to be centralising, is it always going to fall into requiring large centralised power, or does it go somewhere where it becomes something anybody could develop and use and share and develop in a free way? It's an interesting discussion but looking at the last twenty years that's not what's happened. The scientific field and industrial field has become controlled by large pharmaceuticals and agrochemical companies.'

This is undoubtedly true: public sector projects I have worked on such as *Bt* brinjal in Bangladesh or WEMA drought-resistant maize in East Africa are tiny compared to the overall weight of mass-market commodity crops such as genetically engineered soy, maize and canola. So was Jim prepared to draw a distinction between these two things, say *Bt* brinjal in Bangladesh and Roundup Ready soy in Brazil? 'Maybe, maybe. It's an interesting question. I don't know enough about it,' he said. 'But what's clear is that the first generation was all around large monoculture productions in North America that then got exported as an idea.'

Our conversation was interrupted by my son Tom saying goodbye as he went off to play after-school tennis. After he had cycled away up the road, Jim resumed: 'So it's not like saying a technology will always go a certain way, I don't think it's as hardline as that, but I do think there's more likely a sort of a drift towards certain outcomes and uses. And in technology, there's this idea of, there's a latency within a technology.' As he put it in a 2009 speech about synthetic biology: 'We live in an unjust world, and if we're introducing a powerful technology into that unjust world we're probably going to exacerbate that injustice, unless we're very, very, deliberate in trying to attack that injustice. I don't see that happening now.'[15]

Now it was my daughter's turn to interrupt our conversation. 'Dad, don't you know it's time for my dance class? Hurry up!' Rosa does not like being late for her dance class; in fact she likes to be at least 20 minutes early, just to keep us on our toes, as it were. As the father of two children himself, Jim was very understanding. We parted on good terms. I can't speak for how he felt, but I remember feeling both relieved and curiously optimistic. The man I had once been friends with, and then later decided was an ogre ... well, suddenly it seemed like we could almost be friends again. There was a lot we might still disagree on, but I could see that he was a sophisticated thinker, that his motivations were decent and honourable, and that he was doing important, valuable work. And, somewhat disorientatingly, I also remembered how much I liked him.

The problem with being a technology critic, however, is where to draw the line. It is tempting to draw it in the recent past, perhaps in our childhoods, when thanks to nostalgia, things always seem to have been gentler, slower and more familiar. But then this kind of judgement is purely subjective, always dependent on our personal experiences. Even an old typewriter would once have seemed like a fearsomely threatening machine to a calligrapher with a quill.

In an email to me recently Paul Kingsnorth told me that even though he found some of what I had told him about GMOs persuasive, 'I am still against GM ... To me it is a spiritual, and an ethical matter. I just think there are lines we should not cross. I accept that we have already crossed hundreds of them, which is why we're in this mess. But we don't get out of it by crossing hundreds more.' But any attempt at consistency is doomed to fail: Paul was sending me an email, typed on a computer, all courtesy of modern industrial society. And he is resolutely pessimistic about any latter-day attempts to 'stop progress':

> *To me it seems that technology itself has a momentum which is beyond any choices we make. Ultimately, if we – humans – can do something and it profits us, we will do it. The whole of the economic and technological and cultural system is set up for that to be the case. In this context, the Greens are a small protest group that can sometimes hold back the tide, but not for very long. So I see everything we do as operating within this reality: a reality which is inexorably taking us towards the disaster that is already underway. When I talk about spirituality or the sacred in this context, I don't think of it as something that could ever change this momentum or stop us hitting a wall. But I do think it's something we can hold onto, like carrying a small flame in a storm and trying to prevent it from going out. Eventually the storm will pass, though not in our lifetime, and maybe the flame will still be burning. I find that's the best I can do!*

I actually don't share Paul's overall scepticism about industrial society, though only for my own subjective personal reasons. In order to deliver our son Tom in March 2005 my wife Maria had to undergo an emergency caesarean in our fortunately very well-equipped Oxford hospital, the John Radcliffe. Without this modern medical intervention, there is every chance they would both, mother and son, have died during the birth, as happened all too frequently in the quite recent past. That then counts out our daughter Rosa, who came two years later. I contracted pneumonia in 2011, and was in a rapid terminal decline until I was saved by the intravenous infusion of a massive dose of antibiotics, kindly administered by doctors at a hospital in Hereford, where we happened to be at the time. So, in pre-industrial society the entire Lynas family would now be extinct. Call me biased, but I would consider that a disaster unparalleled in the history of the universe, and it can't help but colour my perspective on the goods and bads of the modern world. For me this tension was one of the motivations underlying ecomodernism, as an attempt to try to reconcile the undeniable benefits of industrial society with its equally undeniable drawbacks by using the tools and knowledge of modernity to save the natural world.

Some technology critics, like the poet and farmer Wendell Berry, refuse even to use computers to type on. Berry uses a pencil and paper, writing in an essay unambiguously titled 'Why I am NOT going to buy a computer': 'As a farmer, I do almost all of my work with horses. As a writer, I work with a pencil or a pen and a piece of paper. My wife types my work on a Royal standard typewriter bought new in 1956 and as good now as it was then.'[16] The essay was published in *Harper's* in 1987, sparking several critical letters. The most memorable one was about the implied role of Berry's wife in his righteous crusade: 'Wendell Berry provides writers enslaved by the computer with a handy alternative: Wife – a low-tech energy-saving device. Drop a pile of handwritten notes on Wife and you get back a finished manuscript, edited while it was typed. What computer can do that? Wife meets all of Berry's uncompromising standards for technological innovation: she's cheap, repairable, near home, and good for the family structure.'

This was perhaps unkindly sarcastic but I think the writer made a good point: the political economy concerns of technology point backwards as well as forwards. New tools freed many people, especially women, from the drudgery of unpaid domestic labour. One of my favourite YouTube videos is of a TED talk given by the late Swedish educator Hans Rosling called 'Hans Rosling and the Magic Washing Machine'. In it he tells movingly how the ability to wash clothes in a machine freed his grandmother for the first time to start reading books.

On the other hand, Kirkpatrick Sale, the leader of a movement called the 'Neo-Luddites', is about as hardline as they come. He once gave a graphic demonstration of this anti-technology stance on stage. As he told an interviewer later:

> I was on the stage of New York City's Town Hall with an audience of 1,500 people. I was behind a lectern, and in front of the lectern was this computer. And I gave a very short, minute-and-a-half description of what was wrong with the technosphere, how it was destroying the biosphere. And then I walked over and I got this very powerful sledgehammer and smashed the screen with one blow and smashed the keyboard with another blow. It felt wonderful. The sound it made, the spewing of the undoubtedly poisonous insides into the spotlight, the dust that hung in the air … some in the audience applauded. I bowed and returned to my chair.[17]

Although this sounds like a gimmick, Sale was merely being true to his principles. These are uncompromising. Sale believes that industrial civilisation will collapse soon because of its destruction of the environment and the contradictions between the need for constant growth and disappearing jobs and resources. Sale puts forward handicrafts and social arrangements within tribal societies as better options. But eschewing technology entirely is impossible: humans are a tool-making species, and even he has to admit to his interviewer that he does possess a credit card. And a car. 'One makes accommodations, unless one wants to try to live alone, in the woods. So anybody who wants to stay engaged in the world will have to make some accommodations. The question, I think, becomes, which ones do you make? A lot of Neo-Luddites and techno-resisters today, I think, have made bad choices by saying that they can use the tools of the masters in order to free the slaves. And I don't think this is possible.'[18]

Sale is the author of the superb and very readable book *Rebels Against the Future*, which tells the story of the historical Luddites, the machine-breaking movement that erupted in eighteenth-century England during the early decades of the Industrial Revolution. Jim Thomas too states proudly that he is 'a big fan of the Luddites'. Although popular culture casts the Luddites as ignorant fools resisting the forces of progress, more accurate historical accounts (like Sale's) record that they were not uniformly anti-technology; they opposed machine looms that they saw, quite accurately, as threatening their livelihoods as skilled cottage weavers, and more broadly 'Machinery hurtful to the Commonality'. Sometimes these mechanical looms were smashed in the night, at other times they were dragged to the market place where all the community might gather to pass judgement. Jim Thomas tells of how he was once involved in organising a Luddite-style technology trial, located at York Castle, infamous as the location where 15 men identified as Luddites were hanged in 1812 for the newly established crime of machine-breaking.[19]

'A group of fellow activists dragged a motor car to the old stone tower and we set up public court, inviting bystanders to testify for or against the impact of the internal combustion engine on all our lives. Road kill, asthma, community destruction and climate change were weighed against the increased mobility and economic opportunities provided by four fast wheels. Everyone who happened to pass by became the jury.' The car was found guilty of harming the public good, but, as Jim acknowledged, 'this symbolic exercise in popular assessment of technology was exactly 100 years too late to influence the relevant innovation policy'.[20] However, the experience set him thinking: 'What if we weren't too late? What if we could drag emerging technologies into a modern court of public deliberation and democratic oversight? What might that look like?' It might look something like the worldwide campaign of opposition to genetic engineering – a new technology which many people also saw as harmful, and not always without good cause.

CHAPTER NINE

How Environmentalists Think

So should there be any limits on genetic engineering? Of course, just as there should be limits on any human interference with the world around us. Exactly where these limits should be established is a moral rather than a scientific question. It is mostly a question of intent rather than of explanation – and science is much better at the latter than the former. I've not heard anyone seriously propose that we should undertake human germline genetic engineering, with the single exception of getting rid of nasty and/or fatal genetic diseases. There certainly seems to be no immediate prospect of Jeremy Rifkin's oft-warned slippery slope towards Aryan-selecting eugenics. We should all remain vigilant for sure, but the reason is no doubt because most decent people find the idea that anyone should be allowed to genetically select children for blonde hair and blue eyes morally outrageous. To allow this would be transgressing on a sacred value that our whole society generally upholds, one respecting the inherent worth of a human life.

This sense of upholding sacred values extends to inanimate objects too: who among us would blast apart El Capitan, the sheer glacier-carved mountain in Yosemite, even if a billion-dollar gold seam were accidentally discovered glinting within the granite? When I read about the felling of old-growth forests or the killing of whales, I feel the same sense of moral outrage that something so inherently precious should be so wantonly destroyed. The sense of veneration which most of us share extends also to valued human traditions and ancient monuments. I remember feeling almost physical revulsion when I read about how the terrorist group ISIS had destroyed ancient ruins in Palmyra in Syria by blowing them up with explosives. That anyone could intentionally and callously destroy part of the heritage of all humankind seemed degrading to our collective humanity, just as are war crimes and genocides. But then terrorism depends for its psychological potency on breaching widely held moral norms in the most outrageous way, for example, by targeting innocents such as schoolchildren or people attending music events, or indeed by destroying irreplaceable ancient artefacts.

How far we should respect natural boundaries like those between species is at the heart of the moral debate about genetic engineering. Recall how my friend Paul Kingsnorth told me that 'I just think there are lines we should not cross', fully aware that this was not a rational, scientific or even a consistent position; he admitted that we have already 'crossed hundreds' of such lines, and he was a beneficiary himself of us having done so. Nevertheless to him this was an 'ethical' and a 'spiritual' matter. I think the issue here is about valuing the 'integrity' of what already exists – whether from the natural world or ancient peoples – and not wanting it to be apparently degraded with further human alteration. Personally I am prepared to see more of such lines crossed than Paul is, but not for flippant or unnecessary purposes. I do not believe that there will always be a slippery slope, and that therefore we should block innovations that might have beneficial purposes for some people who need them. I accept however that there is a sanctity about the natural world, and like Paul I do not want human intrusion or alteration to be so ubiquitous that there is nothing left of nature.

The late evolutionary biologist Stephen Jay Gould made this point elegantly in an essay he wrote in 1985, called 'Integrity and Mr Rifkin'. This was, as the title suggests, mostly about criticising Jeremy Rifkin's position on genetic engineering. 'Our world would become a bleak place if we treated living things as no more than separable sequences of information, available for disarticulation and recombination in any order that pleased human whim,' Gould wrote. But, he argued, instead of addressing real fears, opponents like Rifkin had chosen 'an extremism that would outlaw both humane and fascinating scientific research ... I do not see why we should reject all genetic engineering because its

thinking about it was very powerful. I don't mind nearly so much that with an incredibly powerful telescope you can apparently see the marks left on the moon's surface by the Apollo landings. They were undertaken in the spirit of exploration, and added to the proper sense of awe and reverence we feel when contemplating our small place in the universe. Even if the result is the same – a human disturbance on the otherwise pristine surface of the moon – the issue of intent (and of scale) makes all the moral difference in the world.

Sometimes genetic engineering is proposed as a direct aid to the natural world, often to restore the damage done by earlier human intrusion. This might be an example of Kirkpatrick Sale's 'using the tools of the masters to free the slave', and here the moral arguments can cut both ways. One example is the ongoing development of blight-resistant genetically engineered American chestnut trees to replace the chestnut forests that were completely eradicated after a novel pathogen was introduced with imported Japanese chestnut trees in the late nineteenth century. The American chestnut once composed a quarter of all trees in these forests, growing to 100 feet tall and producing nuts consumed by countless animals from squirrels to bears. A team of scientists based at New York State University's College of Environmental Science and Forestry aims to restore this majestic tree's rightful place as a keystone species in the broadleaved forests of the north-eastern United States. The researchers have accordingly spliced a gene from wheat into the chestnut genome that enables it to survive infection by blight.[3]

I can't think of many rational arguments against this. It is highly unlikely that the wheat gene, which occurs in many other grasses, will have any damaging effect on the environment. No squirrels will be poisoned by the resulting GMO chestnuts. All the introduced gene does is allow the tree to produce the enzyme oxalate oxidase, to detoxify the oxalic acid produced by the pathogen which would otherwise kill the tree. All these components are perfectly natural, and their combination could restore vanished ecosystems. All the same, I can't help feeling – yes, feeling – that there is a real qualitative difference between a forest of genetically engineered trees and a forest of natural trees. This doesn't make me oppose the project, but it does give me doubts. Walking through a forest is not the same if you know the trees around you contain a gene that was originally spliced into their genomes in a laboratory. Even if produced for the most altruistic and ecological of purposes, the trees are a partly human artefact, and thus qualitatively different. In some ways they would be a merging of the technosphere and the biosphere. I feel the same way about a wood where people planted the trees (perhaps betrayed by them growing in suspiciously straight lines) as compared to one where the trees arrived and germinated by themselves. My favourite kind of woodland is genuine old-growth, which has been wherever it grows since time immemorial, even before the coming of humans. That connects us to the past, and a sense of transcendence you get from stepping outside the crafted world of modern Man.

These qualitative concerns could still be raised even if the plant or animal is physically and biologically identical to its natural antecedent. My friend Stewart Brand, along with his wife Ryan Phelan, is working on a project called Revive & Restore, which among other things aims to bring back woolly mammoths, passenger pigeons and other extinct charismatic species by extracting DNA from preserved specimens and recreating their genomes artificially. Once again, I can see the point, and even partly share Stewart and Ryan's excitement, but I also have qualms. These animals, should they ever become a reality, might be biologically exact replicas of their vanished ancestors, but they would also be entirely human creations. That makes them qualitatively different.* I

---

* Actually the woolly mammoth would not be a perfect replica as it would be developed by inserting genes for mammoth traits into the genome of its close relative the still-living Asian elephant.

challenged Stewart about this over breakfast once, at a meeting we were both attending in Cavallo Point near Sausalito, in the shadow of the Golden Gate Bridge. He paused with his muesli, gave me a quizzical look, and muttered something about 'essentialism'. He was justifiably pointing out, in other words, that there was no scientifically defensible reason for my scepticism – it was a moral and intuitive reaction.

If the revived woolly mammoth is not wholly 'natural', that begs the question of what constitutes 'nature'. Paul Kingsnorth describes it well in his essay 'In the Black Chamber', where he ponders the question of what 'sacred' means in today's world.

> I realise that what I call 'nature' (an imperfect word, but I can never seem to find a better one) is really just another word for life; an ever-turning wheel of blood and shit and death and rebirth. Nature is fatal as often as it is beautiful, and sometimes it is both at once. But for me, that's the point: it is the fear and the violence inherent in wild nature, as much as the beauty and the peace, that inspires in me the impulses which religions ask me to direct towards their human-shaped gods: humility, a sense of smallness, sometimes a fear, usually a desire to be part of something bigger than me and my kind.

It is this essence of something beyond the merely human that would for both me and Paul be lacking in a human-altered or recreated species.

Not surprisingly, Paul also has serious doubts about Revive & Restore. Although he admits that he can 'see where Brand's idea has come from', Paul's instinctive reaction was 'horror'. He writes cuttingly: 'Say what you like about religion, but at least it teaches us that we are not gods. The ethic that is promoted by the de-extinctors and their kind tells us that we are gods and we should act like them.' This, in Paul's view, is nothing less than 'the latest expression of human chauvinism; another manifestation of the empire of *Homo sapiens*' that must ultimately lead to the 'end of the wild', a 'New Nature, entirely the product of our human-ness'. He ends the passage with this zinger:

> *'I am Stewart Brand, reviver of extinct species', declaims Brand on the web forum Reddit. I am Ozymandias, king of kings: pleased to meet you.*

Paul is at pains in his essay to defend what he regards as his legitimately emotional reaction to the Revive & Restore project, without needing to resort to the usual utilitarian 'valid criticisms' in order to make his case. Anticipating that some – not least Brand – might call his response 'irrational', he admits: 'Very well, then: it is irrational, and it is no less real for that.' He writes: 'Humans have changed much, and we control much, or try to, but we have never stepped over this threshold before. We have never moved towards the creation of life itself, and the consequent, inevitable elimination of wild nature.'

Artificial mammoths aside, is genetic engineering in agriculture inherently stepping over this new threshold? I guess it is, but then so – as Paul admitted in his email to me – have been all the other thousands of incremental innovations that preceded it, from the selective breeding that developed today's staple food crops through to the invention of the plough. Even our non-GMO crops, as domesticated species, are already a long way from their natural ancestors. Google 'teosinte' to see what the original wild ancestor of maize looked like: it is a bushy plant with seed spears carrying just a few hard seeds arranged in an unappealing row at the end. Modern wheat is even more unnatural: *Triticum aestivum* resulted from a three-way hybrid between original wild Einkorn wheat and two different species of a weed called goat grass. The wheat that goes into our daily bread is thus a hexaploid, containing six sets of chromosomes (humans and most – but by no means all – natural species are diploid, with two sets of chromosomes, one from each parent).[4] Wild bananas are almost inedible, containing round hard seeds that break your teeth. Cultivated bananas are seedless and therefore they are all cloned as suckers from parent plants, making them genetically identical. The same goes for potatoes, which are cloned via transplanted tubers rather than suckers.

Selective breeding of crops depends on either hybridisation or fortuitous mutations. These may make the resulting plant yield more fruit, harvest more easily or taste better, for example. But this selective breeding process typically takes decades, even centuries. One way to speed it up is to increase the rate of mutations by introducing a mutagen, either a potent toxic chemical or a source of ionising radiation, either of which will break apart DNA in numerous places. The resulting mutations will mostly be negative, harming the plant and resulting in damage. But every now and then something useful happens – some varieties of durum wheat for pasta, ruby red grapefruit and 'Golden Promise' barley were created in this way using mutagenesis. Interestingly, mutagenic crops are included in organic systems: they are not subject to the same ban as GMOs. Today a debate is raging about whether point mutations (a mutation or change affecting only one base-pair of DNA) using precise genome editing tools such as Crispr should be considered GMOs or not. Crispr is like a pair of molecular scissors that can snip DNA at exactly the sequence targeted. Excluding it while allowing radiation mutagenesis makes no sense at all; it is like telling a brain surgeon they can use a blunderbuss but not a scalpel.

Cultivated crops also need constant and intense human involvement in order not to die out. You might find weeds invading a wheat field, but you will never find wheat invading a weed field (or a woodland). This is one reason why I do not mind genetic engineering of crop plants, because this will have little effect on neighbouring ecosystems (except to the extent to which the introduced genes allow changes in farming systems, such as different levels of pesticide use). But on a more moral level, I do not object to the genetic engineering of crop plants because these are already largely human creations. My concerns stem more about technology interfering with nature in a more direct sense. I am more worried about having GMOs in my local woodland than I am about having them in my food. The latter is fine – as biologists have long argued, all our cultivated plants have been extensively genetically modified already.

The only argument that can really be raised against GMOs in general is the 'unknown unknowns' concern, that there might be some inherent danger or damage to the genome due to the introduction of recombinant DNA. It is very difficult to imagine how this could be the case. DNA is broken apart and repaired all the time within living cells – sunlight, metabolic oxygen and countless other insults create damage in DNA often tens of thousands of times per cell per day. These are quickly repaired by cellular enzymes. Every time a cell divides, at least 10 double-strand DNA breaks happen, on average.[5] Mostly, enzymes join up nearby broken ends in the DNA molecule, but they can also assimilate new genetic sequences that happen to be available. So genetic engineering is not much different to the kind of thing cells have evolved to do constantly already. It is vanishingly improbable that there is some inherent risk that applies to genes introduced by genetic engineers in the lab and not to genes mutated and repaired naturally countless times within the cell.

Nevertheless, genetic engineering feels different because it is seen to be unnatural, and this leads to the widespread intuition that it must somehow be inherently risky. You can see this in the language used: GMOs might 'contaminate' organic crops, causing 'genetic pollution', but never the other way round. Professor Séralini's cancer-riddled rats worked brilliantly at this emotional level. No one needed to read his text (which was indecipherable Frenglish anyway) and no amount of complaining by his many critics that Séralini's stats were shoddy, that the controls also got cancer, that the rats' diets weren't equivalent, and so on, made any impact against the sheer gut horror of the colour photos he published showing rats with hideous outsized tumours. Statistical significance be damned when emotions this strong are involved. Entire countries were persuaded in an instant; as I showed earlier, Séralini's rats led to a GMO ban in Kenya that continues to this day. Séralini's paper, even though it was retracted, is constantly cited around the world as evidence of the harm that GMOs surely cause.

Although I defend the right of anyone to have a fundamental moral objection to genetic engineering, I find Kenya's ban objectionable because it offends my equally strong commitment to empiricism. I don't want to see pseudoscience leading to policy change in Kenya or anywhere. I would much rather that the implicit moral objections to GMOs be made explicit and the debates around them conducted accordingly ('you can't put a bacterial gene in this maize because I believe that to do so is wrong'). I was glad when the European Union bowed to the inevitable and allowed its member states to cite broad 'public policy' (that is, moral) reasons for wanting to ban GMO cultivation on their territories.[6] This is far better than torturing and twisting 'scientific evidence' to try to fit a preconceived position, as the French and other European governments had been doing for years. It is not shameful to reject scientific evidence when it conflicts with a moral case, so long as this is done explicitly. It is not fine to twist scientific evidence so that it can be used as a rationalist fig leaf to obscure an implicit moral case. If objections are part of the moral realm, let them be discussed as such – all of us, genetic engineers included, believe that there should be limits on how this technology is used, and that these should be ethically defined. So let's have an honest debate about where the limits should be set.

I also suspect that some of the ostensibly political objections to GMOs emanate from this same sense of moral transgression. As I suggested to Jim Thomas during our Skype call, many of the same concerns about corporate control, centralisation, monopolisation and so on could apply equally – more, in fact – to the iPhone than to genetically modified soy. Although Jim told me, with creditable consistency, that he did have concerns about mobile phones too, it is undeniable that communications technologies are not nearly as emotionally charged as food and farming. Thus we are particularly alert for moral outrages when we feel that something unnatural and transgressive has already been done to the genetics of our beloved food crops. Seen this way, the controversies over India's *Bt* cotton farmers and Séralini's rats seem much more explicable. This also gives a moral underpinning to objections that patenting is an 'enclosure' of the public property of the genome. I have to say I share these objections. I find the idea that anyone could 'own' any of my genes offensive. In most countries the law is on my side here: in 2013 the US Supreme Court declared that naturally occurring DNA sequences were indeed ineligible for patents.

One example of genetic engineering that I do cautiously support is the transgenic salmon project called AquaBounty, which provocatively brands itself as 'the world's most sustainable salmon'. While there is a place for wild salmon fishing, humanity's appetite for delicate pink sashimi and succulent salmon fillets is already far too high to be satisfied from wild sources alone. Hence the need for aquaculture, which after all is just doing in the seas what we adopted millennia ago on land in moving from hunter-gathering to farming domesticated animals in pens. However the environmental problems of salmon farming at sea are well known: nutrient pollution can accumulate and affect marine ecosystems; sea lice can also contaminate wild salmon, as can the anti-lice chemicals used to remove them; and fish can escape to interbreed with wild relatives. In addition, salmon are carnivores, so they are fed with fish meal harvested from the wild oceans, and this in effect just pushes the pressure of overfishing down the food chain.

AquaBounty's model is very different. Its fish are Atlantic salmon with an added growth hormone gene from Chinook salmon, which keeps the fish growing all year round. This means that they require less feed to reach full market weight, reducing the pressure on wild fish stocks. Genetic engineering could also help to replace fish meal entirely with land-based feedstocks – scientists have already experimentally succeeded in developing oilseed crops that express the omega-3 fatty acids that currently can only be derived from oceanic protein sources. This has not persuaded anti-GMO groups like the Center for Food Safety to support transgenic salmon, however. CFS has launched lawsuits to block the approval of AquaBounty salmon on the basis that 'any approval of GE fish would represent a fatal threat to the survival of native fish populations' and that 'the environmental threats posed by GE fish could alter the biodiversity of entire ecosystems'.[7]

I don't find this convincing, because AquaBounty's genetically engineered salmon are not going to be anywhere near the sea. The company is instead intending to raise them only in land-based tanks far away from coastal wild salmon populations. This is better for the environment, they claim, because effluent can be captured and used as fertiliser, while water can also be recycled.[8] An additional safeguard is that all the fish raised will be sterile females. In order for cross-breeding with wild salmon to occur, therefore, the GE fish would have to jump out of their tanks, sprout legs, haul themselves all the way to the coast, swim to wherever salmon populations are, change sex and cease to be sterile at the same time. Even the most technologically adept 'Frankenfish' are unlikely to be capable of such feats. But what these absurdities illustrate is that the debate is not about the specifics of AquaBounty – it is about a moral objection that CFS holds to genetic engineering in principle. I can understand this, but I think that the goal of a truly sustainable source of salmon makes it a step worth taking – though only if the salmon is treated as a domesticated species a long way from its wild setting. If GE salmon were to be raised in sea pens alongside wild fish, I would be happy to sign the Center for Food Safety's petition.

All these different GMO controversies show that we ignore or displace people's sense of morality at our peril in what are construed as only scientific debates. In his excellent book *The Righteous Mind*, the social psychologist Jonathan Haidt shows that people will go to almost any lengths to justify their emotional responses to a moral transgression in 'acceptable' rationalist language. In experiments to investigate this phenomenon, Haidt asked people about situations which tended to invoke disgust but which had actually not harmed anyone. One example was, to paraphrase, 'the family dog was knocked down and killed by a car – the family then sat down to eat its dead flesh. Nobody saw them.' Most people feel a shudder of revulsion at the disrespectful treatment of the dog's corpse, but struggle to justify this in rational terms on the basis of any definable harm. In order to justify their moral intuitions people therefore tended to invent victims: they told Haidt that the family might be harmed by getting sick from eating dog meat, for example, or that a neighbour might have been offended by seeing the family eat roast dog (ignoring the explicit statement in his story that no one had seen the act). Haidt concluded that 'it was obvious that most of these supposed harms were post hoc fabrications. People usually condemned the actions very quickly – they didn't seem to need much time to decide what they thought. But it often took them a while to come up with a victim, and they usually offered those victims up halfheartedly and almost apologetically.'[9]

Haidt reports that many of these post hoc justifications for what were in reality instinctive, immediate and reflexive moral judgements were so unconvincing as to be almost preposterous. What they had in common was that they all tried to find some external evidence of harm to explain their disgust about something that they clearly felt was *essentially* wrong. When Haidt gently debunked each argument in turn, the less confident or argumentatively able among them might even run out of excuses altogether and become what he termed 'morally dumbfounded', saying things like 'I know it's wrong, but I just can't think of a reason why.' As Haidt concluded: 'These subjects were reasoning. They were working quite hard at reasoning. But it was not reasoning in search of truth; it was reasoning in support of their emotional reactions.'[10] The moral convictions of the interviewees even held when they were repeatedly proven wrong: 'they seemed to be flailing around, throwing out reason after reason', Haidt recalled.

None of this is intended to suggest that people with moral intuitions are somehow irrational or silly. They are simply being human, and Haidt was trying to find out how humans tend to behave and why. Indeed it is people *without* moral intuitions who cause the most serious problems – they are called psychopaths. Haidt's conclusion is sobering, however. Forget, he writes, about ever persuading anybody of anything through reasoned evidence and force of argument. 'If you ask people to believe something that violates their intuitions, they will devote their efforts to finding an escape hatch – a reason to doubt your argument or conclusion. They will almost always succeed,' he says.[11] Haidt thereby refutes the 'deficit model' of human psychology. This is the idea that many people have – including, ironically, most scientists in my experience – that everyone who disagrees with them is simply ignorant, and that presenting them with a barrage of scientific 'facts' will educate them sufficiently to change their opinions in line with one's own. I suffer from this myself; I have never found that telling anyone that the rate of farmer suicides in India is no higher than in France or Scotland has the slightest effect on their beliefs about the rightness or wrongness of GMOs. But I keep on doing it regardless, including in this book.

The suicides story is powerful because it offers an external justification for an intuitive moral rejection: a victim, and the more miserable and exploited the victim (poverty-stricken Indian farmer), and the more evil the villain (Monsanto) the better. To go back to Haidt's initial hypothetical morality tale, imagine how much more outraged you might feel if you were told that a rich businessman had eaten a poor family's dog, and that the children whose much-loved puppy this once was had seen him doing it and wept inconsolably. This kind of moral framework is almost impossible to break with rational analysis – the best that can happen is that it is inverted and replaced with an equally powerful but alternative moral framework that one hopes is more reflective of the actual truth. I have tried to do something like this in my chapter on

Africa, where the anti-GMO groups are the villains and the farmers are the victims because they have been denied the right to choose better seeds that might help them emerge from poverty. The anti-GMO groups working in Africa have the opposite narrative of course. They think that farmers who buy 'improved' seeds will be exploited by multinationals, undermining local resilience and food sovereignty. Note that you can flip these narratives over, rather like the famous Necker cube – a line-drawn transparent cube you can see two ways – but as with the Necker cube you can't see or believe both at the same time.

As my chapter on Africa shows, it is not just anti-GMO campaigners who are motivated by moral concerns. Pro-GMO campaigners are equally passionate, and as a result can be just as deaf to reasoned argument that cuts across their biases. One example is the Greenpeace co-founder Patrick Moore, who left Greenpeace a few years after its founding, and more recently disavowed many of the issues the group stands for in a 2010 book entitled *Confessions of a Greenpeace Dropout*. Moore has set up a campaign called 'Allow Golden Rice Now!' that has charged Greenpeace with 'crimes against humanity' for opposing genetically modified Golden Rice (a bio-fortified rice intended to tackle vitamin A deficiency, which particularly affects young children in developing countries). 'We believe that Greenpeace's actions during the past 14 years to prevent Golden Rice from being produced and reaching the millions of people who now suffer needlessly from vitamin A deficiency constitute a crime against humanity as defined by the Rome Statute,' declares Moore's website.[12] That is a moral position for sure. However, Moore is on thin ice claiming that Greenpeace has singularly deprived malnourished children from receiving life-saving bio-fortified rice. Although Greenpeace has indeed campaigned against it, and it could be convincingly argued that its campaign has contributed to a negative regulatory and political environment that has harmed the project, so far – and despite decades of technical work – the team at the International Rice Research Institute has not yet put forward Golden Rice for full cultivation approval in any country*. So it is not obviously true that Greenpeace has 'blocked' Golden Rice, as many pro-GMO campaigners have claimed.

Even the most celebrated scientists in the world – people who, having received Nobel Prizes in their fields, might be expected to behave with utmost rationality at all times – betray their flawed humanity in their own Golden Rice campaign. Although it is written in scientific language, and bolstered by the signatures of no less than 124 Nobel laureates, the June 2016 'Letter to the Leaders of Greenpeace, the United Nations and Governments around the world' was an intensely moral rather than merely scientific statement. Calling on Greenpeace 'to cease and desist in its campaign against Golden Rice specifically, and crops and foods improved through biotechnology in general', the statement insisted that 'opposition based on emotion and dogma contradicted by data must be stopped', and ended by asking: 'How many poor people in the world must die before we consider this a "crime against humanity"?'[13] There can hardly be a more emotionally loaded approach than to accuse people you disagree with of 'crimes against humanity', employing the sort of language that is usually reserved for the instigators of wars and genocides. Yet here more than a hundred of the finest scientific minds in the world were letting their moral intuitions so far cloud their rational judgement that they could sign a statement making claims that were only thinly supported by actual evidence. So scientists are human after all – who knew?

It is true that scientists may sometimes (though all too infrequently in my experience) be persuaded by empirical evidence to change their minds. Richard Dawkins tells a nice story in his book *Unweaving the Rainbow*, which he uses to demonstrate the uniqueness of the scientific method in challenging human bias. He remembers as one of the

---

* Update: Golden Rice was put forward for regulatory approval in the Philippines in August 2017.

'formative experiences' of his Oxford undergraduate years when a visiting lecturer from the United States presented new scientific evidence that 'conclusively disproved the pet theory of a deeply respected elder statesman of our zoology department, the theory that we had all been brought up on'. Instead of angrily challenging this new evidence, 'the old man rose, strode to the front of the hall, shook the American warmly by the hand and declared, in ringing emotional tones, "My dear fellow, I wish to thank you. I have been wrong these fifteen years."' As Dawkins asks rhetorically, and with feeling, 'Is any other profession so generous towards its admitted mistakes?'[14]

Dawkins's example is perhaps an unusual situation because it comes from a debate between scientists – in other words, an environment that is socially constructed specifically to give primacy to evidence and reasoned argument. This is very much the exception rather than the rule, and even more so when debates revolve around morally loaded issues, whether these involve GM food or gun control.

This is also why I still strongly defend the importance of the scientific consensus on the safety of genetically engineered foods. Opponents insist that statements by institutions such as the Royal Society or the American Academy for the Advancement of Science are mere evidence of groupthink on the part of the scientists. This is not just the case with the GMO debate; opponents of the scientific mainstream on issues like climate change and vaccinations make similar arguments alleging collective bias on the part of the scientific community. I don't think this makes sense. The flawed Nobel laureates' statement is not, to me, evidence that a scientific consensus is meaningless; these experts were signing something that was mostly outside their areas of genuine expertise (and it is always a mistake to imagine that someone who is an expert in something is an expert in everything). On the other hand, the Intergovernmental Panel on Climate Change has a meticulous and time-consuming process to review the scientific literature and formulate a consensus of expert opinion on global warming. The US National

Academy of Sciences recently undertook a similar process on genetic engineering, producing a voluminous report that once again 'found no substantiated evidence that foods from GE crops were less safe than foods from non-GE crops'. I optimistically wrote a blog post in response entitled 'The GMO safety debate is over'.[15] I was wrong of course.

So why does no one believe even the National Academy of Sciences? Why do people hold so firmly to their ideas even when hundreds of world experts disagree with them? At the very basic level, no one wants to be wrong. Admitting error is always difficult – when blamed for making a mistake most of us instinctively find reasons why we are innocent, were acting with the best of intentions, or some other excuse to limit the psychological pain and imagined reputational damage. This is one reason why the scientific method is so counter-intuitive: science advances through errors, specifically requiring the disproving of previous theories. Only when a theory resists multiple attempts at falsification should a hypothesis be declared – temporarily at least – to be objectively true. Hence the 'null hypothesis', the basis of statistical evidence test. The starting point for examining statistics is the hypothesis that nothing is happening. Only when so-called 'p values' (probability values) are very low (usually below 0.05), and therefore the chance of a result coming out a certain way purely by chance is very unlikely (below 5 per cent), is a result considered 'statistically significant'. And even then this should be the starting point rather than the final conclusion, not least because out of every 100 scientific papers declaring statistical significance for their results, five of them will be wrong by pure chance under the very same 5 per cent probability test.

Whereas in science being wrong is an integral part of advancing knowledge, in politics admitting to being wrong is the kiss of death for anyone aspiring to leadership. In the UK, BBC Radio 4 has a morning news show called *Today*, where the objective of every interview with every politician is to back him or her into a rhetorical corner in order to force an admission of the dreaded 'U-turn', which is considered a knockout. One of the most famous examples of the political perils of being seen to change your mind is the vilification of the Democratic presidential candidate John Kerry during the 2004 US election. After changing his mind on key issues like the Iraq war, Kerry was accused of 'flip-flopping', talking 'waffle' and being 'wishy-washy' – all of which did his campaign immense damage. As Kathryn Schultz writes in her book *Being Wrong*: 'Jay Leno proposed two possible slogans for the Kerry campaign: "A mind is a terrible thing to make up," and "Undecided voters – I'm just like you!"' During the Republican Convention, delegates took to doing a kind of side-to-side stadium wave whenever Kerry's name was mentioned: a visual waffle (or call it a waver). For ten bucks, you could purchase a pair of actual flip-flops – the footwear, I mean – with Kerry's face on them.'[16] The result of all this of course was that the wise and infallible George W. Bush was returned to power.

Most of us, not being political leaders, are able to hold happily deluded views of ourselves as being generally righteous and moral without much danger of these delusions being exposed for what they truly are. Perhaps this is necessary for our mental health – those who see themselves as no better than anyone else and are constantly racked by doubt are also likely to be first in the queue for anti-depressants. However, in the absence of moral constraints, imposed primarily by our awareness of the judgements of others, most of us do in fact lie and cheat much of the time, and some of us would commit or at best ignore far worse crimes. We may even deny it to ourselves, but countless experiments have shown that these are enduring components of typical human behaviour. As Schultz reminds us: 'I think of this as the French Resistance fantasy. We would all like to believe that, had we lived in France during World War II, we would have been among those heroic souls fighting the Nazi occupation and helping ferry the persecuted to safety. The reality, though, is that only about 2 percent of French citizens actively participated in the Resistance. Maybe you and I would have been among them, but the odds are not on our side.'[17]

This hints at the reality, which is that our morality and behavioural standards are largely a product of our social conditioning, which is itself very much dependent on specific cultural and historical situations. More than anything – more than our tool-making, our opposable thumbs or our cerebral cortex – humans are a social species. For us, group dynamics are critical to every aspect of our behaviour. As Jonathan Haidt explains, people don't care about evidence in political or moral questions; that is why people who lack health insurance are not any more likely to vote Democrat. What people care about is their groups, 'whether racial, regional, religious or political' (or, more likely, several at the same time). Thus: 'Political opinions function as "badges of social membership." They're like the array of bumper stickers people put on their cars showing the political causes, universities, and sports teams they support. Our politics is groupish, not selfish.'[18]

Perhaps the most critical insight that Haidt shares in his book is that our reasoning faculties have not evolved to help us search for anything resembling objective reality: they have evolved to help us be more valued and successful members of our groups. For a social species like humans, being ejected or excluded from a group would be a probable death sentence, thus adding powerful evolutionary selective pressure for groupish behaviour. Haidt writes that our reasoning 'evolved not to help us find truth but to help us engage in arguments, persuasion, and manipulation in the context of discussions with other people.'[19] This is why confirmation bias is so powerful, because it is 'a built-in feature (of an argumentative mind), not a bug that can be removed (from a platonic mind)'.

When I read Haidt's assertion that 'conscious reasoning is carried out largely for the purpose of persuasion, rather than discovery', this immediately made me reconsider my own experience. The story I had been telling myself was that I reasoned my way out of anti-GMO beliefs through discovering and assimilating scientific information that I had earlier been lacking. As I remembered shamefully, I hadn't even known what DNA stood for during my old crop-trashing days. This seemed to make me a case study for the 'deficit model', where new information can change someone's mind by addressing their lack of knowledge. And yet I also knew that I had had plenty of opportunities to learn more about why GMOs might not be so bad back in my anti-GMO days, and I had not been in the least bit interested in taking those opportunities. When I debated with pro-GMO scientists in the media or at events it was not to learn more about their perspective; it was to defeat them, the more conclusively the better. So far as I was concerned, it was the scientists who were narrow-minded, not me. Years later I was asked by one of these same experts, a genetics professor at a college in Oxford, if there was anything he could have said differently at the time to convince me. I told him I didn't think so. It wasn't that their arguments lacked force. Their mistake was to think that their arguments mattered much at all.

Within groups there are powerful forces that act to prevent people from changing their minds and challenging group consensus. The common-sense term for this is 'peer pressure', though we fondly tend to imagine that this applies only to schoolchildren rather than to all of us as human beings. Kathryn Schultz calls this a community's 'disagreement deficit', a kind of groupthink. Schultz writes: 'First, our communities expose us to disproportionate support for our own ideas. Second, they shield us from the disagreement of outsiders. Third, they cause us to disregard whatever outside disagreement we do encounter. Finally, they quash the development of disagreement from within. These factors create a kind of societal counterpart to cognition's confirmation bias, and they provoke the same problem. Whatever the other virtues of our communities, they are dangerously effective at bolstering our conviction that we are right and shielding us from the possibility that we are wrong.'[20] Confirming my response to the Oxford genetics professor, Schultz explains: 'Even when we do encounter outside challenges to our beliefs, we usually disregard them. In fact, much as we tend to automatically accept information from people we trust, we tend to automatically reject information from unfamiliar or disagreeable sources.'

Groupthink can be damaging when it goes too far because it excludes contrary opinions that might be correct and useful, meaning that groups might behave sub-optimally. A particularly infamous example is a historical one: the Salem witch trials, where a whole community seemed to become possessed by a kind of hysteria driven by powerful conformity pressures that led to 19 people being hanged for witchcraft on the evidence of nothing more than dreams and visions. The psychologist Irving Janis defined groupthink as 'a mode of thinking that people engage in when they are deeply involved in a cohesive in-group, when the members' strivings for unanimity override their motivation to realistically appraise alternative courses of action'.[21] This certainly describes my experience as an environmental activist. We had a number of unwritten but strongly policed rules, such as that no one should criticise 'the movement' externally, especially from within what we derisively termed the 'mainstream media'.

This kind of groupthink can be repressive and exclusionary, but is also vital for a community to maintain cohesion and address a combined problem. As activists we would never have been able to take the radical measures we did, risking violence, prison or worse to try to stop the bulldozers moving in to destroy a piece of ancient woodland or hillside, without very tightly knit, loyal groups. Sometimes these were formally constituted and termed 'affinity groups', where members were expected to look out for each other whatever the circumstances and personal cost. Challenging group consensus may occasionally have been useful and necessary, but constant internal opposition could not be tolerated for long. As Schultz writes, 'a lone dissident can destroy the cohesiveness of an entire community. From this latter perspective, doubt and dissent represent a kind of contagion, capable of spreading and destroying the health of the communal body. Accordingly, many communities act quickly to cure, quarantine, or expel (or, in extreme cases, eliminate) any nonconformists among them.'[22]

This certainly explains the vigour with which religions often persecute non-believers, with especial hatred reserved for those who have abandoned their faith. Schultz tells the story of a Muslim man in Afghanistan who converted to Christianity and was forced to flee the country. I saw this at first hand too in the Maldives (where I was climate advisor to the president from 2009 to 2012). The Maldives claims to be a '100% Muslim country', where alcohol is banned and your bags are checked at the airport for the illicit importation of forbidden articles like Bibles and bacon. On one occasion a Maldivian man admitted to atheism, meaning that the country could no longer claim to be 100 per cent Muslim. The main tenor of the resulting national debate was not whether or not the dissenter should be punished, but what kind of execution would be appropriate: stoning or beheading? He was forced to appear on television, recant and beg for forgiveness.[23] Members of a Maldivian Facebook group advocating secularism were later abducted by vigilantes and threatened with death unless they recited verses from the Koran.[24] President Nasheed, whom I worked for, tried to pursue a more secular, human rights-respecting policy, but even he was powerless when religious passions were inflamed in the general public.

I do not mean to compare my friends from the environmental movement with religious extremists. The only thing these cases have in common is that they are examples of very human groupish behaviour. This is something that all of us do, all of the time. Every time we share a post on Facebook about a political subject, we are advertising our group loyalty to our friends. As Jonathan Haidt writes, our reputations are just as important to us as food and shelter – probably more so because, in evolutionary terms, membership of a group was essential in order to secure food and shelter at all. This kind of behaviour does not make us unthinking or bad. It makes us human.

All this rings true to me because I remember how I felt when I began to challenge group consensus in the environmental movement. This was not only about GMOs – I also did it with nuclear power, and at other times during gatherings and meetings. Once Paul Kingsnorth and I wrote anonymous satirical nursery rhymes about radical activism which were printed and distributed at an EarthFirst! gathering in Wales.[25] Paul was not there, but I was quickly found out and given the full pariah treatment. I remember how strongly the feeling reminded me of being back at school and being excluded from some playground game where everyone would run away from you and hide. Even though I was well aware that I had brought it on myself, it was still a disorienting experience. I accepted my self-imposed exclusion and spent the time climbing nearby mountains instead.

This happened in about the year 2000, well before I began to have a change of heart on GMOs. So the truth about my experience was not that I found out scientific information challenging my position on GMOs and then changed my mind accordingly, forcing my expulsion from the environmental movement. The truth was that I had already begun to pull away from at least the direct action wing of the environmental scene many years earlier, for numerous other reasons. Partly this was about feeling jaded and cynical, but more importantly, I knew I wasn't very good at direct action and wanted instead to become a writer. When I later spoke and wrote publicly about my change of mind on GMOs, therefore, I could risk the resulting storm of criticism from environmentalists because I no longer identified so strongly with that group.

Reading Haidt's book on group dynamics made me understand that I was probably primed to change my mind on GMOs only because I had begun to shift my loyalty from one group, the greens, to another, the scientists. Receiving the Royal Society Science Books Prize in 2008 I took, rightly or wrongly, as a trophy of affirmation from the scientific community. If I'd been a tribal headhunter this would have been the equivalent of bringing back the scalp of an enemy

chief. And it was only when my reputation was threatened –
because my writing on GMOs was shown to be perilously
unscientific by the sorts of people I now felt aligned with –
that I had to seriously reconsider my position.

In other words, deep down I probably cared less about
actual truth than I did about my reputation for truth within
my new scientific tribe. Hopefully these ended up being
more or less the same thing. But this all happened only
because I was in a fairly unusual position, as a science writer
with a reputation for scientific accuracy to maintain. It wasn't
so much that I changed my mind, in other words. It was that
I changed my tribe.

CHAPTER TEN

Twenty Years of Failure

In November 2015 Greenpeace published a report entitled *Twenty Years of Failure: Why GM crops have failed to deliver on their promises*. 'Despite twenty years of pro-GM marketing by powerful industry lobbies, GM technology has only been taken up by a handful of countries, for a handful of crops,' the report states. It adds that only 3 per cent of the world's agricultural land is used to grow genetically modified crops, and that the vast majority of this GM crop acreage is devoted only to two traits: herbicide tolerance and insect resistance. It also notes that European consumers 'do not consume GM foods' and that a single type of maize is the only GM crop cultivated in the whole continent, concluding triumphantly: 'Most of Asia is GM-free, with the GM acreage in India and China mostly accounted for by a non-food crop: cotton. Only three countries in Africa grow any GM crops. Put simply, GM crops are not "feeding the world".'

The only part of this I would dispute is the claim that only 3 per cent of the world's agricultural land is used to grow GMO crops. Actually the latest figure is 12 per cent of global cropland, including about half of all US arable land.[1] The rest of it is true enough. What the report's writers fail to note, though, is that what they perceive as an inherent failure of genetic engineering technology might be at least partly explained by the very success of Greenpeace and others in blocking it. In a striking piece of circular reasoning, Greenpeace is in effect saying: 'This technology we have been fighting to stop for the last 20 years hasn't been very successful yet.' Starting in 1996, as I showed earlier, Greenpeace led a global campaign to hamper the deployment and development of genetically modified crops. It spent the subsequent two decades fighting virtually every single introduction of a GMO anywhere in the world. Now, in a textbook example of a self-fulfilling prophecy, Greenpeace writes a report entitled *Twenty Years of Failure*, claiming that GMOs have not delivered on the promises once made for them. QED. It might as well be Greenpeace's annual report.

I have presented numerous examples already in this book showing how successful the anti-GMO movement has been in blocking adoption of this technology. As Greenpeace notes, whole areas of the world have been shut off from the promised benefits of biotech, and only a few commodity crops have been deployed successfully with genetically engineered versions. Not only have whole crop types – wheat, potato, rice – been directly blocked by activist campaigns and wider public fear, but also costs for the whole technology have skyrocketed because of the need to compile voluminous safety dossiers for regulators in every country, and the long waiting time – from years to decades – needed to get new seeds to market. This means that only the most profitable mass-market global commodity crops have been worth investing in for biotech companies, while huge numbers of once-promising ideas and developments have been mothballed or ditched altogether.

Non-corporate actors such as public sector plant research agencies and academic institutions cannot afford to get their ideas past regulators, so promising innovations sit on the university laboratory shelf too. It is not worth anyone's time to spend years developing GE crops for the small horticultural market. So the argument that many campaigners make that genetic engineering only benefits big corporations and big farmers is another circular one. Activism has been most successful in locking out small and public sector players from the biotechnology revolution, thus cementing exactly the monopolistic situation that many campaigners say they are fighting against.

One of the few examples of a public sector genetic engineering project that actually came to fruition – literally – was the development of virus-resistant papaya. This saved the family-farming papaya industry in Hawaii, which was being devastated by the papaya ringspot virus in the late 1990s.

Fortunately a team led by researchers at Cornell University managed to create a resistant variety using a coat protein gene taken from the virus itself. The new 'Rainbow' papaya has performed well ever since. It is also, I can report from personal experience, delicious. However, in recent years there has been an upsurge in anti-GMO activism in Hawaii, with ballot initiatives to ban all genetic engineering on the islands, boycott campaigns against the virus-resistant papaya, and vandalism sometimes seeing whole orchards on family farms hacked down with machetes in the night.[2]

Originally virus-resistant papaya was also planned for Thailand, which was similarly affected by ringspot virus, and where papaya is an important part of people's diets and food culture. In this case, Thai government researchers teamed up with Cornell University to introduce the virus resistance gene in Thai varieties of papaya, famed for *som tam* green salad. Field trials were then established in Thailand. However, these performed so well that in 2004 news soon spread and seeds were spirited away for use by local farmers in advance of official approval for their release. Spotting an opportunity, Greenpeace put out press materials claiming that this was 'one of the worst cases of genetic contamination of a major food crop in Asia' and even including a map of 'the spread of contamination in Thailand'.[3] Greenpeace activists also scaled the fence into the government research station, chopped down some of the test papaya trees and put up a banner reading 'Stop GMO field trials'. Embarrassed, the Thai government quickly acceded to this demand, forcing the researchers to destroy the rest of their experiment and bury the trees in pits on the site.[4] Officials then scoured the countryside to remove and destroy any genetically engineered papaya trees. Ironically these could sometimes be easily identified as the only ones in the vicinity not affected by the disease. Not surprisingly the Greenpeace action meant that some Thai farmers experienced hardship and loss of livelihoods as a result of the destruction of what has been termed a 'pro-poor' GM crop.[5]

Crucially, the furore also had a chilling effect on the whole plant biotech sector in Thailand. Crops in development at that time included chilli and tomato resistant to viral diseases, yard-long bean and cotton resistant to insect pests, virus-resistant and salt-tolerant rice, and numerous others. All were shelved: the success of Greenpeace in turning virus-resistant papaya into a national 'contamination' scandal killed off Thailand's budding biotech industry virtually overnight. Nearly 15 years later, there is still no officially approved transgenic papaya in Thailand, nor are any other GMO crops being grown or even in development. Of a reported 40 or so projects underway in the early 2000s, every single one was discontinued.

Although every country is different, similar campaigns have succeeded in blocking GMO crops in numerous locations around the globe. Greenpeace and local allies blocked *Bt* aubergines in the Philippines via both vandalism and court action, as I mentioned earlier. In India anti-GMO groups succeeded in wrestling an indefinite moratorium out of the spineless national government in 2010. This has yet to be formally rescinded. Not a single Indian GM crop has been approved since that time; the public sector developers of GM mustard are currently fighting for approval for their crop, but seem likely to lose owing to the ferocity of the campaign against them.

I have already related numerous stories of how activists have blocked GM crop projects in Africa. In South America, Peru has an official 10-year moratorium in place. Ecuador, Venezuela and Chile also forbid widespread cultivation. In all of Europe the situation is pretty much beyond retrieval; Hungary even wrote an anti-GMO clause into its national constitution. The Russian government threatens severe fines for either importers or cultivators of GMOs. The issue seems to span geographies and politics unlike almost any other. On a trip to China recently I heard how – after Greenpeace succeeded in fomenting a national scandal over a Golden Rice feeding trial to a group of young children in 2012[6] – GMOs are now seen by millions of people as an American plot to poison the nation's children.

So it is indeed true, as Greenpeace states, that GM crops are not significantly important in 'feeding the world'. But that is down in no small part to the efforts of Greenpeace and its allies in stopping them from doing so.

So what might a fairer summary of the overall impacts of genetically modified crops be? A 2014 meta-analysis, combining the outputs of nearly 150 separate peer-reviewed studies, concluded that GM technology adoption has reduced chemical pesticide use by 37 per cent, increased crop yields by 22 per cent and increased farmer profits by 68 per cent globally.[7] Most of these farmers, incidentally, are in developing countries. This global aggregate picture of course lumps together a huge amount of data, some positive, some negative. But it should at least give green groups pause, I would have thought, that the technology they have been opposing for 20 years has reduced chemical pesticide use by 37 per cent. Certainly I had no idea when I first began to campaign against Monsanto and Roundup Ready that GM would actually *reduce* the use of chemicals in farming (almost all these reductions come from insecticides reduced by the *Bt* trait; herbicide tolerance has mostly just changed the types of herbicides used, shifting most world use towards glyphosate).

There are other environmental issues too. One global study estimated that the worldwide adoption of GM crops led to savings in 2015 of about 26 million tonnes of carbon dioxide, thanks to fewer insecticide sprays and the soil carbon storage improvements associated with no-till GM crops.[8] This is equivalent to taking 12 million cars off the road for the year. It sounds a lot, but actually is not a huge deal in global terms: 26 million tonnes is roughly what might be released each year by seven large coal-fired power stations.[9] The US alone has a fleet of over 600 coal-fired stations, and China more than 2,000.[10] If your overriding priority is the climate therefore, getting coal off the grid is still a far higher priority than pushing GM crops. (Greenpeace has a Quit Coal campaign which I am happy to endorse here.)[11] But the takeaway is clear – in climate terms, GM crops do also come in broadly positive.

This is not necessarily the case for other environmental considerations. Although early suggestions that pollen from *Bt* corn might harm North America's famed monarch butterfly populations were not supported by later studies, there is evidence that the herbicide tolerance trait has led to a significant decline in the monarch's main food plant, milkweed, in several corn- and soy-growing US states.[12] Many other factors affect the population of monarchs, from deforestation in their overwintering grounds in Mexico to the vagaries of the weather. But all other things remaining equal, a reduction in their food source cannot be good for the resilience of one of North America's iconic species. In response Environmental Defense Fund (EDF) is now working with farmers across the US in a 'Monarch Butterfly Habitat Exchange' in order to provide them with incentives to protect milkweeds. 'Since farmers and ranchers manage much of the habitat appropriate for milkweed, they are in a perfect position to restore and enhance this vital habitat, creating key corridors of breeding and nectaring habitats along the monarch butterfly's great migration,' EDF says.[13] EDF's effort is one of several, all of which are only needed of course because US farmers are now so successful in controlling weeds, partly thanks to herbicide-tolerant GMOs.

On the other hand, the adoption of *Bt* crops has been good for insect biodiversity, because of the associated reduction of insecticide sprays. One study from China found that reduced pesticide spraying led to higher populations of beneficial insects such as ladybirds, lacewings and spiders on *Bt* as compared to non-*Bt* cotton.[14] On the other hand, overuse of glyphosate has certainly sped up the evolution of weeds that are resistant to this herbicide to 35 at the last count, including the fabled cornfield 'superweed', Palmer amaranth. Herbicide-resistant weeds eventually evolve anywhere herbicides are used often enough, GMO or non-GMO: France has its quota of glyphosate-resistant weeds too, although it does not grow any genetically engineered crops.

Although gene flow has occurred – the much-feared 'contamination' of natural ecosystems – most of this comprises a few feral glyphosate-tolerant canola, alfalfa or bentgrass volunteers springing up along roadsides and along irrigation canals and ditches. There is no sign of the *Bt* gene moving outside cultivated plants on farms. The National Academy of Sciences (NAS) reports as an official 'finding': 'Although gene flow has occurred, no examples have demonstrated an adverse environmental effect of gene flow from a GE crop to a wild, related plant species.'[15] Superweeds? Contamination? Both these things have happened – but they have no meaningful environmental impacts at all. They are trivial. Environmentalists can stop worrying about them, and get on with addressing more serious issues.

Worryingly, Greenpeace's *Twenty Years of Failure* document states that it is a 'myth' that 'GM crops are safe to eat'. It reports that 'there is no scientific consensus on the safety of GM foods', supporting this claim with a reference to a 'joint statement' produced in 2015 by 'over 300 independent researchers'. This statement, which was produced and signed by many of the leading anti-GMO activist researchers and campaigners around the world, was headed 'No scientific consensus on GMO safety' and published in a low-ranking journal called *Environmental Sciences Europe*.[16] 'A broad community of independent scientific researchers and scholars challenges recent claims of a consensus over the safety of genetically modified organisms,' these dissenting voices say in their paper's abstract. 'Claims of consensus on the safety of GMOs are not supported by an objective analysis of the refereed literature.'

Reading this paper reminds me of similar 'no consensus' statements produced by dissenters in other areas of scientific controversy, such as climate change, vaccines or HIV/AIDS.

For example, the promoters of one climate denialist petition claim to have gathered 31,000 scientific signatures – a hundred times more than the GMO sceptics managed – to state that 'there is no convincing scientific evidence that human release of carbon dioxide will, in the foreseeable future, cause catastrophic heating of the Earth's atmosphere and disruption of the Earth's climate'.[17] In 2001, in response to a PBS (Public Broadcasting Service) documentary on Darwinian evolution, a Creationist think-tank gathered 100 scientific names to declare that they were 'skeptical of claims for the ability of random mutation and natural selection to account for the complexity of life'. The signatories included impressive-sounding people such as a professor of cellular and molecular physiology at Yale, as well as various mathematicians, biologists and physics experts from top-ranked universities.[18] Going by the formal academic credentials of the signatories, this was also a more impressive list than the anti-GMO one.

I cite these two examples to show that self-selected groups of dissenters do not really challenge scientific consensus, even when they include a few impressive-sounding names. There are at least half a dozen climatologists I can think of who categorically deny human-caused global warming, just as there are famous virologists based at prestigious universities who insist that HIV does not cause AIDS. Consensus does not mean 100 per cent agreement, nor is it a battle between whoever can compile the longest and most impressive-looking list of experts. Perhaps the best illustration of the perils of self-selected group statements is the National Center for Science Education's 'Project Steve', which is 'a tongue-in-cheek parody of a long-standing creationist tradition of amassing lists of "scientists who doubt evolution" or "scientists who dissent from Darwinism".' Project Steve now has a list of over 1,400 signatories supporting Darwin's theory of evolution, all scientists named Steve.[19]

What these no consensus efforts serve to do in reality is give spurious pseudoscientific justification for those who want to deny the existence of scientific consensus for ideological reasons, whether religious, political or something else. I find it troubling to see Greenpeace so obstinately on the wrong side of world scientific opinion on this issue. This is not just because I admire Greenpeace as an organisation for its other environmental work, but also because this ideological selectivity undermines the credibility of its other campaigns, which it also claims are backed up by scientific evidence. If Greenpeace denies science on GMOs, how do we know it is true then what Greenpeace says about overfishing? Or deforestation? Or biodiversity? Or even climate? The organisation is on the horns of the same dilemma I once was: you cannot defend one scientific consensus while denying the other and still expect to be trusted on issues of science. It really is that simple, and it is high time for Greenpeace to recognise it.

Lest I be accused in turn of cherry-picking, let me quote in detail from the aforementioned National Academy of Sciences 2016 report on GMOs. The full document including appendices also considers socio-economic aspects of GMOs, but since it runs to 388 pages in total I will confine myself to the simpler safety question here. Interestingly the report begins by noting that the scientific consensus on genetic engineering has shifted dramatically over the decades. In 1974 an NAS committee chaired by Nobel laureate Paul Berg warned that there was 'serious concern that some of these artificial recombinant DNA molecules could prove biologically hazardous'. This was around the time, as I described in an earlier chapter, that environmental groups first assembled in force to draw attention to this potential hazard. They were justified in doing so according to the limited science of the time, when very little was known about potential risks. However, by 1987, after more than a decade of further evidence had been assembled and considered, the NAS had changed its mind. An NAS committee concluded that 'the risks associated with the introduction of R[ecombinant]-DNA-engineered organisms are the same as those associated with the introduction of unmodified organisms and organisms modified by other methods' and that such organisms posed no unique environmental hazards. Further NAS reports issued in

1989, 2000, 2002 and 2004 gradually reinforced this position. None of them found any adverse health effects in humans attributable to genetically engineered crops.

The NAS could change its mind because as a scientific institution it was mandated to take an evidence-based position on the issue of recombinant DNA, and to continually reassess its position as scientific evidence might change over time. It has a formal process to review the scientific literature and to adjust its conclusions accordingly. It is instructive that the environmental groups did not behave similarly, having no such formal process for ensuring evidence-based policy-making. As political interest groups, once their position was formulated it became ideologically more rigid over time even as the scientific evidence underpinning it became weaker and weaker. They were unable to adapt their position in the light of changing evidence, just as politicians are unable to admit to doing U-turns, partly because they feared losing status and credibility in the process. Anti-GMO campaigning also became an interest, with professional reputations and salaries built on it. The Center for Food Safety, with its multi-million dollar turnover, would threaten its very existence if it took a science-based position on GMOs. But most importantly, a moral narrative – a negative mental frame – about genetic engineering had been constructed, which proved resilient in the face of any challenge from scientific realities.

These days the more extreme claims about cancer, autism and the like are not typically made openly by mainstream groups. They are however increasingly ubiquitous on the internet, and are aggressively promoted by fringe groups like the Organic Consumers Association, alternative medicine peddlers such as Mercola.com, and by individuals like Jeffrey Smith who do not have scientific credentials. Smith, for instance, is a former yogic flyer and dance instructor, who studied at the Maharishi School for Management, part of the Maharishi movement that promotes 'Transcendental Meditation'.[20] The spectacular rise of the organic movement has also driven many people to seek out 'healthy' alternatives to GMOs, which are identified with conventional agriculture and assumed to be heavy on chemicals and environmental impact. These themes are staples of the alt-right too – on his Infowars channel conspiracy theorist Alex Jones treats his viewers to occasional tirades against 'poisonous' GMOs and glyphosate. These combined efforts have been incredibly successful: opinion surveys show that roughly 40 per cent of Americans – evenly divided, by the way, between Republicans and Democrats – believe that foods with genetically modified ingredients are 'worse for health' than conventional foods.[21] In a celebrated 2015 study the Pew Research Center (a non-partisan think-tank and polling organisation based in Washington DC) found that there was a wider gap between the general public and the scientific community on GMOs than on any other area of equivalent controversy, from vaccines to evolution to nuclear power.[22]

So what do the facts say? In its report the NAS publishes a graph showing cancer incidence – the lines go up and down for different cancer types, but there is no change from 1996 when GE foods were first introduced. The unsurprising finding is that 'the data do not support the assertion that cancer rates have increased because of consumption of products of GE crops'. Moreover, 'patterns of change in cancer incidence in the United States are generally similar to those in the United Kingdom and Europe, where diets contain much lower amounts of food derived from GE crops'. Furthermore, contrary to popular wisdom, cancer deaths have actually declined in the US and Canada in recent decades.

Nor is there any data linking GE foods with kidney disease. How about obesity? Or diabetes? 'The committee found no published evidence to support the hypothesis that the consumption of GE foods has caused higher U.S. rates of obesity or type II diabetes.' The same conclusion holds for coeliac disease, various allergies and autism. For the latter, there has been a rapid increase in diagnoses in recent decades, but at the same rate in the US and UK. And these are on the negative side of the ledger: on the positive side, genetic engineering can be used to increase beneficial nutrients (such as vitamin A) and reduce toxins, like the possible carcinogen acrylamide in fried potatoes.

I am not conflating the safety issue with other more legitimate concerns about GMOs. There are numerous political, social and economic issues about corporate concentration, small farmers, herbicide use and so on that continue to be relevant and of concern to many critics of the technology. However, I think we can now be clear that food safety objections to GMOs are scientifically untenable, even if they are a potent worldwide rabble-rousing tool. The ends do not justify the means here, however. As ActionAid found out to its cost when it was exposed for running adverts on Ugandan radio claiming that GMOs cause cancer, environmental and development groups run a serious reputational risk if they spread pseudoscientific scare stories about GMOs. Our most respected campaigning groups should not be purveyors of post-truth. The debate on GMOs will continue on numerous other grounds – whether political, economic, moral or spiritual – but opposition should not be based on outright lies about 'gay genes', cancer and autism.

Meanwhile, denying a worldwide scientific consensus requires extreme selective bias. This is the ultimate in cherry-picking. Greenpeace highlights a statement by a small group of dissenters, while ignoring the National Academy of Sciences, the American Association for the Advancement of Science, the Royal Society, the African Academy of Sciences, the European Academies of Science Advisory Council, the French Academy of Science, the American Medical Association, the Union of German Academies of Science and Humanities and numerous others. Even the European Commission admitted in a 2010 report: 'The main conclusion to be drawn from the efforts of more than 130 research projects, covering a period of more than 25 years of research, and involving more than 500 independent research groups, is that biotechnology, and in particular GMOs, are not per se more risky than e.g. conventional plant breeding technologies.'[23]

Greenpeace also selectively quotes from the scientific literature, misrepresenting the positions of reputable international institutions in order to bolster its case. Its *Twenty Years of Failure* report quotes the World Health Organisation making the following statement: 'Different GM organisms include different genes inserted in different ways. This means that individual GM foods and their safety should be assessed on a case-by-case basis and that it is not possible to make general statements on the safety of all GM foods.' This appears to challenge the case for scientific consensus on GMO safety, doesn't it? No, because Greenpeace did not quote the full statement. The very next sentence from WHO reads as follows: 'GM foods currently available on the international market have passed safety assessments and are not likely to present risks for human health. In addition, no effects on human health have been shown as a result of the consumption of such foods by the general population in the countries where they have been approved.' Whoops! This kind of selective quotation is embarrassing – it shows just how weak Greenpeace's case must be that its only option in trying to claim institutional support is to blatantly misrepresent a UN body.

Changing an individual mind is difficult, changing a collective organisational position doubly so. But Greenpeace would not even need to be the first. Recently one of the earliest groups opposing genetic engineering, the Environmental Defense Fund, revised its position on biotechnology after lengthy internal debate. The resulting statement reads: 'EDF recognizes the use of biotechnology as a legitimate deployment of science in the search for effective solutions, and also recognizes that past deployment of some biotechnology products has caused legitimate concerns. For that reason, we will support or oppose specific biotechnology products or processes based on transparent assessments of their health, environmental, social, and economic risks and benefits. The risks and benefits of biotechnology products will often vary by organism, geography and other variables, and need to be assessed at relevant temporal and spatial scales. New products and techniques resulting from the application of biotechnology, like products of all science, technology, and engineering, need to be evaluated for their risks and benefits, including social implications, before they are deployed. For this reason, EDF does not support or oppose broad categories of biotechnology products, such as genetically modified organisms (GMOs), and recognizes that some proposed products may not result in beneficial outcomes or warrant support.'[24]

Come on Greenpeace, follow EDF and accept the science – is this really such a terrifying prospect?

What if Greenpeace won and GMOs were banned? This was the hypothetical case investigated by a team from Purdue University in the US. 'This is not an argument to keep or lose GMOs,' Professor Wally Tyner explained about the study in a press release. 'It's just a simple question: what happens if they go away?'[25] The Purdue team's agricultural model suggested that eliminating all GMOs in the United States would lead to corn yield declines of 11 per cent, cotton yield declines of 18 per cent and losses in soy production of 5 per cent. Making up for these losses in US genetically engineered corn, cotton and soy yield by growing non-GE crops instead would lead to an increase in cropland area of about 1.1 million hectares, the Purdue team estimated, with about a third of that being forest land, roughly 380,000 hectares. 'This means that adopting GMO technology avoided conversion of natural land (forest and pasture) to cropland,' the Purdue team concluded.[26]

How important is this land-sparing effect of GMOs? According to the UN Food and Agriculture Organisation, the world sees a net loss of 3.3 million hectares of forest on average every year,[27] so the area spared by the higher crop yields delivered courtesy of GMOs is only about a tenth of what is destroyed annually. This is not a huge amount – but then the Purdue study only concerns US GMO production, about 70 million hectares out of a global total of 180 million. Via a back-of-the-envelope calculation, assuming the same proportional yield losses would apply globally, I estimate forest losses of just under a million hectares in a worldwide GMO-free scenario. That is about half the area of Wales, or a little less than the US state of Connecticut, or just under a third of the annual area of net global forest loss. Not a huge deal perhaps, but not completely trivial either. I wouldn't personally want to be pushing for an area of forest half the size of Wales to be cut down unless there was a very good reason.

Perhaps the most interesting result emerging from the Purdue study didn't actually involve GMOs at all. The team also looked at the land losses resulting from the US ethanol programme, which saw about 15 billion gallons of ethanol produced in 2016,[28] gobbling up fully 40 per cent of US corn production.[29] The Purdue team found from their model that the loss of forest and other natural land resulting from this enormous diversion of corn into biofuels almost exactly matched the losses they would expect to see if US-grown GMOs were banned. Because of these land-use impacts and deforestation, environmental groups are rightly sceptical of biofuels. As Greenpeace itself explains: 'When land used for food or feed production is turned over to growing biofuel crops, agriculture has to expand elsewhere. This often results in new deforestation and destruction of other ecosystems, particularly in tropical regions in the developing world.'[30] Quite right. However, if Purdue's modelling study is correct, Greenpeace's two policies cancel each other out: banning GMOs would almost exactly negate any forest savings resulting from getting rid of the US corn ethanol programme.

The obvious better plan is to take the best of both options by getting rid of biofuels and keeping GMOs, thereby sparing a whole Wales-worth of forest from destruction. For me this illustrates how the GMOs debate needs to move away from black and white either/or framings and towards a more nuanced yes/and/but/maybe approach. According to many studies we are now well into what has been termed the Earth's 'sixth mass extinction' because of human impacts on biodiversity. In order to spare as many species as possible, we have to protect as much land (and ocean) area as we can. The ecologist and conservationist E. O. Wilson has proposed that a target of half the Earth's land area should be set aside for wildlife.[31] In order to achieve this, there must be a drastic reduction in per-capita consumption and a continuing decline in human population growth. Wilson acknowledges that high technology can make a contribution, pointing out that 'food production per hectare [is] sharply raised by indoor vertical gardens with LED lighting, genetically engineered crops and microorganisms', among other innovations. All land is not the same: Wilson is clear that protection needs to focus on high-value ecosystems such as the redwood forests of California, the Amazon river basin, the cloud forests of the Andes, the Galapagos islands and the Congo basin.

George Monbiot, a proponent of large-scale rewilding, proposes that vanished species be reintroduced to increasingly large protected areas. All of this, however, will require the maintenance of high-yield crops on existing (though hopefully reducing) areas of farmland. These need not be sterile deserts: even high-yield agriculture, small or large scale, can ensure that wildlife is supported and encouraged as far as possible. Once again this is not either/or. As George writes in his excellent book *Feral*: 'While I would argue against a mass rewilding of high-grade farmland, because of the threat this could present to global food supplies, we lose little by allowing nature to persist in small fallow corners and unexploited pockets of even the most fertile places.'[32] George points out that the food/wilderness trade-off looks less daunting if you start with marginal areas that are already very inefficient in producing food. His prime example from the UK is the extensive upland area devoted to raising sheep, which makes a negligible contribution to the UK's meat supply while suppressing the biodiversity value of vast sweeps of mountain, moorland and hillside landscape.

Reducing global meat consumption could be the most significant factor of all in protecting natural ecosystems. George Monbiot is hence a reluctant vegan, writing in the *Guardian*: 'Rainforests, savannahs, wetlands, magnificent wildlife can live alongside us, but not alongside our current diet.' I saw the potential of global vegetarianism – or better, veganism – in the Global Calculator model I worked on with the late David MacKay and his colleagues (you can find this model on globalcalulator.org). It presents numerous different pathways on everything from dietary choices to transport patterns to power-generating sources. The idea is to keep global temperatures – represented by a little red thermometer in the top right-hand corner – from rising above two degrees. It is very difficult to do, but the task becomes almost easy if you reduce global meat consumption down to the average of most Indians. There is also an option of improving crop yields, which helps too. On the other hand, if you select options where the whole world has a diet as meat-intensive as the United States, you get a little error message: 'Your pathway uses more land than the world has available. Please change your settings!' High beef consumption scenarios can even break the model's little thermometer, and by implication the global climate.

So let's hear it for the GMO proponents. But let's also hear it for the vegans, the conservationists, the farmers, the scientists, the environmentalists and indeed everyone who is working to understand how we can best protect this planet both for future human generations and for the rest of life around us. Let's use science as the wonderful tool that it is, but let's also respect people's feelings and moral intuitions about the proper extent of human intrusion into the biosphere. Maybe now we can finally join forces to ensure that scientific innovations, in agriculture as much as anything else, are critically assessed and deployed in a way that helps the environment and improves the livelihoods of people in poorer countries too.

Above all, let us not repeat the mistakes of the past. We have already wasted 20 years fighting over a mere seed-breeding technique that – used sensibly and in the public interest – can certainly help global efforts to fight poverty and make agriculture more sustainable. Let's not waste 20 more.

Notes

Chapter 1: UK Direct Action

1 This is described by Jim Thomas in his chapter in Tokar, B. 2001. *Redesigning Life? The Worldwide Challenge to Genetic Engineering.* Zed Books, London.

2 Press release, 7 August 1997. UK Gene Crop Destroyed, www.gene.ch/gentech/1997/Jul-Aug/msg00487.html mentions the High Wycombe occupation.

3 Someone – I don't know who – did talk about this action, and it made the *Daily Record* in 1998. *Daily Record*, 21 June 1998. Good Golly Dolly; Kidnap threat to cloned sheep.

4 Squire, G. R., et al. 2003. On the rationale and interpretation of the farm-scale evaluations of genetically modified herbicide-tolerant crops. *Philosophical Transactions of the Royal Society of London* B 358: 1779–1800.

5 *The Guardian*, 21 September 2000. Greenpeace wins key GM case.

6 *BBC News*, 16 February 1999. GM food taken off school menu, news.bbc.co.uk/1/hi/education/280603.stm.

7 *BBC News*, 8 March 1999. Fast-food outlets turn against GM food, news.bbc.co.uk/1/hi/uk/292829.stm.

8 *Sunday Post-Dispatch*, 25 July 1999. Fear is growing; England is the epicenter.

9 *Sunday Post-Dispatch*, 25 July 1999. The English make it clear to the world that they don't want to mess with Mother Nature.

10 Genetix Update newsletter, Autumn 1999, no. 14. Available from www.togg.org.uk/togg/updates/GUissue14.pdf

11 Jim Thomas, in Tokar, *Redesigning Life?* Ibid., p. 340.

12 *Independent*, 12 July 1999. UK's 'most eco-friendly' trees are destroyed by GM activists.

Chapter 2: Seeds of Science

1   Press release, 5 September 2001. 'Pies for damn lies and statistics' as Danish anti-green author gets his just desserts, www.urban75.com/Action/news138.html.

2   Undercurrents TV, 2001. Bjorn Lomborg pied by Mark Lynas, www.youtube.com/watch?v=TOg8IqkS4PA

3   *The Guardian*, 17 June 2008. Lynas's Six Degrees wins Royal Society award, www.theguardian.com/books/2008/jun/17/news.science.

4   Waltz, E. 2009. GM crops: Battlefield. *Nature*, 461: 27–32.

5   Gilbert, N. 2013. Case studies: A hard look at GM crops. *Nature*, 497: 24–26.

6   *New Statesman*, 30 May 2005. Mark Lynas: Nuclear power – a convert, www.newstatesman.com/node/195308?page=2.

7   Lynas, M. 2010. Why we greens keep getting it wrong, *New Statesman*, www.newstatesman.com/environment/2010/01/nuclear-power-lynas-greens.

8   *The Australian*, 18 January 2013. An inconvenient truth, www.theaustralian.com.au/news/inquirer/an-inconvenient-truth/news-story/0fc19aaf635f9dc97bed3ff538961c9e.

9   *New York Times*, Dot Earth blog, 4 January 2013. New Shade of Green: Stark Shift for Onetime Foe of Genetic Engineering in Crops, dotearth.blogs.nytimes.com/2013/01/04/.

10 *GM Watch*. Background Briefing – Mark Lynas and the GM movement in the UK, gmwatch.org/en/background-briefing-mark-lynas-and-the-gm-movement-in-the-uk.

11 BBC World *HARDTalk*, 30 January 2013.

12 *Rothamsted Research*. 2012. GM Appeal. www.YouTube.com/watch?v=I9scGtf5E3I.

13 Bruce, T., et al. 2015. The first crop plant genetically engineered to release an insect pheromone for defence. *Nature Scientific Reports*, 5: 11183.

Chapter 3: The Inventors of Genetic Engineering

1   Van Beveren, E. Statues – Hither and Thither, www.vanderkrogt.net/statues/object.php?record=beov025&webpage=ST.

2   Schell, J. 1975. The Role of Plasmids in Crown-Gall
    Formation by A. Tumefaciens. In: Ledoux L. (eds) *Genetic
    Manipulations with Plant Material*. NATO Advanced Study
    Institutes Series (Series A: Life Sciences), vol 3. Springer,
    Boston, MA.

3   *WUNC North Caroline Public Radio*, 23 February 2015.
    The Life, Legacy and Science of 'Queen of Agrobacterium'
    Mary-Dell Chilton, wunc.org/post/life-legacy-and-science-
    queen-agrobacterium-mary-dell-chilton#stream/0
    (47 minutes).

4   Van Montagu, M. 2011. It Is a Long Way to GM Agriculture.
    Annual Review of Plant Biology, 62: 1-23.

5   Monsanto, 1997. Fields of Promise: Monsanto and the
    development of agricultural biotechnology.

6   Charles, D. 2001. *Lords of the Harvest: Biotech, Big Money, and
    the Future of Food*. Basic Books, Cambridge, US.

7   Robinson, D. and Medlock, N. 2005. Diamond v.
    Chakrabarty: A Retrospective on 25 Years of Biotech
    Patents, *Intellectual Property & Technology Law Journal*, 17, 10:
    12-15.

8   Robb Fraley, interview by Brian Dick at Monsanto, St Louis,
    Missouri, 16 December 2015 (Philadelphia: Chemical
    Heritage Foundation).

9   Robb Fraley, interview by Brian Dick at Monsanto, Ibid.

10 Charles, D. 2001. Ibid., p. 5.

11 *World Food Prize*, The Sculpture. www.worldfoodprize.org/
    en/about_the_prize/the_sculpture/.

Chapter 4: A True History of Monsanto

1   Forrestal, D. 1977. *Faith, Hope, and $5000: The Story of
    Monsanto: The Trials and Triumphs of the First 75 Years*. Simon
    and Schuster.

2   Myers, R. 2000. *The 100 Most Important Chemical Compounds:
    A Reference Guide*. Greenwood, Westport, US.

3   *International Directory of Company Histories*, 2006. Monsanto,
    www.encyclopedia.com/social-sciences-and-law/

economics-business-and-labor/businesses-and-occupations/
monsanto-company.

4 *Greenfields*. Astroturf, www.greenfields.eu/astroturf/.

5 *Wired*, 6 December 2009. 12 June 1957. Future is now in
   Monsanto's house, www.wired.com/2009/06/
   dayintech_0612/.

6 Institute of Medicine of the National Academies. 2012.
   Veterans and Agent Orange: Update 2012. p. 55.

7 *National Pesticide Information Center*. 2,4-D Technical
   Fact Sheet, npic.orst.edu/factsheets/archive/2,4-DTech.
   html.

8 Institute of Medicine of the National Academies, 2012,
   Ibid.

9 *New York Times*, 19 April 1983. 1965 Memos Show Dow's
   Anxiety on Dioxin.

10 *New York Times*, 1983. Ibid.

11 *New York Times*, 6 July 1983. Ralph Blumenthal: Files Show
   Dioxin Makers Knew of Hazards.

12 *New York Times*, 30 November 1993. Alison Leigh Cowan:
   Veterans Seek Revival of Agent Orange Suit.

13 *New York Times*, 11 March 2005. William Glaberson: Civil
   Lawsuit on Defoliant in Vietnam is Dismissed.

14 Carson, R. 1962. Chapter 2, The Obligation to Endure.
   Silent Spring. Penguin, London.

15 *New York Times Magazine*, 21 September 2012. How 'Silent
   Spring' Ignited the Environmental Movement.

16 *CBS News*, 19 September 2012. The Price of Progress, www.
   cbsnews.com/videos/the-price-of-progress/.

17 Stoll, M. 2012. Industrial and agricultural interests fight back.
   Virtual Exhibitions, Vol. 1, www.environmentandsociety.org/
   exhibitions/silent-spring/.

18 *Monsanto Magazine*, October 1962. The Desolate Year,
   iseethics.files.wordpress.com/2011/12/monsanto-magazine-
   1962-the-desolate-yeart.pdf.

19 *Scientific American*, 4 May 2009. Should DDT Be Used to
   Combat Malaria?

20 *Washington Post*, 1 January 2002. Monsanto Hid Decades of
   Pollution.

21 *New York Times*, 29 February 2016. Chemical Safety Bill Could Help Protect Monsanto Against Legal Claims.

22 *National Geographic*, July 1979. See www.flickr.com/photos/jbcurio/8740859605.

23 *The Atlantic*, 2 December 2014. Bhopal: The World's Worst Industrial Disaster, 30 Years Later.

24 *New York Times*, 30 October 2014. Warren Anderson, 92, Dies; Faced India Plant Disaster.

25 *Holocaust Education & Archive Research Team*. I. G. Farben.

26 *Bloomberg*, 5 February 2015. America's Most Loved and Most Hated Companies.

27 *New Yorker*, 3 November 2013. Why the Climate Corporation sold itself to Monsanto.

28 *New York Times*, 10 June 1990. Betting the farm on biotech.

29 *New York Times*, 10 June 1990. Ibid.

30 Schurman, R. and Munro, W. 2010. *Fighting for the Future of Food: Activists Versus Agribusiness in the Struggle over Biotechnology.* University of Minnesota Press, US, pp. 37–8.

31 Schurman, R. and Munro, W. 2010. Ibid., pp. 43–4.

32 *Industry Task Force on Glyphosate*, 2017. Glyphosate Facts. Glyphosate: mechanism of action, www.glyphosate.eu/glyphosate-mechanism-action.

33 *Monsanto.com*. Monsanto History: An Introduction, www.monsanto.com/whoweare/pages/monsanto-history.aspx.

34 Quoted in Schurman, R. and Munro, W., 2010. Ibid., pp. 33–4.

35 Schurman, R. and Munro, W., 2010. Ibid., p. 133.

36 *CropLife*, 17 July 2017. Complexity in Agriculture: The Rise (and Fall?) of Monsanto, www.croplife.com/management/complexity-in-agriculture-the-rise-and-fall-of-monsanto/.

37 Robb Fraley, interview by Brian Dick at Monsanto, Ibid.

38 *Monsanto.com*, 1 December 2015. Monsanto Takes Action to Fight Climate Change with Carbon Neutral Crop Production Program, monsanto.com/news-releases/monsanto-takes-action-to-fight-climate-change-with-carbon-neutral-crop-production-program/.

39 *CIP*. Biosafety and Health, research.cip.cgiar.org/confluence/ display/potatogene/The+NewLeaf+story.

40 *New Yorker*, 10 April 2000. The Pharmageddon Riddle.

41 Vaeck, M., et al. 1987. Transgenic plants protected from insect attack. *Nature*, 327: 33-37.

Chapter 5: Suicide Seeds?

1 Monsanto.com, 11 April 2017. Myth: Monsanto Sues Farmers when GMOs or GM Seed is Accidentally in Their Fields, monsanto.com/company/media/statements/gmo-contamination-lawsuits/.

2 *Right Livelihood Award*. Percy and Louise Schmeiser, 2007, Canada, www.rightlivelihoodaward.org/laureates/ percy-and-louise-schmeiser/.

3 Percy Schmeiser – David versus Monsanto, www.youtube. com/watch?v=oPKoSrc99p4.

4 Monsanto.com, 11 April 2017. Percy Schmeiser, monsanto. com/company/media/statements/percy-schmeiser/.

5 *MIT Technology Review*, 30 July 2015. As Patents Expire, Farmers Plant Generic GMOs.

6 Monsanto.com, 9 April 2017. Roundup Ready Soybean Patent Expiration, monsanto.com/company/media/ statements/roundup-ready-soybean-patent-expiration/

7 *The Wiglaf Journal*, June 2012. Monsanto & the Global Glyphosate Market: Case Study, www.wiglafjournal.com/ pricing/2012/06/monsanto-the-global-glyphosate-market -case-study/.

8 Supreme Court of the United States, *Bowman v. Monsanto Co. et al*. Decided 13 May 2013, www.supremecourt.gov/ opinions/12pdf/11-796_c07d.pdf.

9 Monbiot, G., 1 January 1997. Science with Scruples – Amnesty Lecture, www.monbiot.com/1997/01/01/ science-with-scruples/.

10 Klümper, W. and Qaim, M. 2014. A Meta-Analysis of the Impacts of Genetically Modified Crops, *PLoS One*, 9, 11: e111629.

NOTES 277

11 *Center for Food Safety & Save Our Seeds*, 2013. Seed Giants vs. US Farmers, www.centerforfoodsafety.org/files/seed-giants_final_04424.pdf.

12 *GMOanswers.com*, 2014, gmoanswers.com/ask/why-does-monsanto-sue-individual-farmers-and-other-ag-biotech-companies-dont-if-they-do-it,

13 Monsanto Fund, Our Mission, www.monsantofund.org/about/our-mission/.

14 Organic Seed Growers and Trade Association, et al, v. Monsanto, www.osgata.org/wp-content/uploads/2011/03/OSGATA-v-Monsanto-Complaint.pdf.

15 *Mother Jones*, 1 December 2012. DOJ Mysteriously Quits Monsanto Antitrust Investigation, www.motherjones.com/food/2012/12/dojs-monsantoseed-industry-investigation-ends-thud/.

16 *Mother Jones*, 1 December 2012. Ibid.

17 *ETC Group*, 15 September 2016. The Monsanto–Bayer tie-up is just one of seven; Mega-Mergers and Big Data Domination Threaten Seeds, Food Security, www.etcgroup.org/content/monsanto-bayer-tie-just-one-seven-mega-mergers-and-big-data-domination-threaten-seeds-food.

18 *Food & Water Watch*, 26 July 2017. American Antitrust Institute, Food & Water Watch, and National Farmers Union Say Monsanto-Bayer Merger Puts Competition, Farmers, and Consumers at Risk, www.foodandwaterwatch.org/news/american-antitrust-institute-food-water-watch-and-national-farmers-union-say-monsanto-bayer.

19 AAI, Food & Water Watch and National Farmers Union, 26 July 2017. Re: Proposed Merger of Monsanto and Bayer, www.foodandwaterwatch.org/sites/default/files/white_paper_monsanto_bayer_7.26.17_f.pdf.

20 *Daily Mail*, 3 November 2008. The GM genocide: Thousands of Indian farmers are committing suicide after using genetically modified crops.

21 Bitter Seeds, teddybearfilms.com/2011/10/01/bitter-seeds-2/.

22 *New Yorker*, 25 August 2014. Seeds of Doubt.

23   Shiva, V. Monsanto vs Indian Farmers, vandanashiva.com/?
     p=402.
24   *New Yorker*, 25 August 2014. Ibid.
25   Kathage, J. and Qaim, M. 2012. Economic impacts and
     impact dynamics of *Bt* (*Bacillus thuringiensis*) cotton in India,
     PNAS, 109, 29: 11652–11656.
26   Krishna, V. and Qaim, M. 2012. *Bt* cotton and sustainability
     of pesticide reductions in India, *Agricultural Systems*, 107:
     47-55.
27   Krishna, V. and Qaim, M. 2012. Ibid.
28   Cornell Alliance for Science, 30 October 2014. *BT* Cotton in
     India – The Farmer's Perspective, allianceforscience.cornell.
     edu/bt-cotton-india-farmers-perspective.
29   Plewis, I., 2014. Indian Farmer Suicides – Is GM cotton to
     blame? *Significance*, Royal Statistical Society.
30   *The Conversation*, 12 March 2014. Hard Evidence: does GM
     cotton lead to farmer suicide in India? theconversation.com/
     hard-evidence-does-gm-cotton-lead-to-farmer-suicide-in-
     india-24045.
31   *The Conversation*, 12 March 2014. Ibid.
32   Plewis, I. 2014. Ibid.
33   Feed the Future South Asia Eggplant Improvement
     Partnership. Pesticide use in Bangladesh, bteggplant.cornell.
     edu/content/facts/pesticide-use-bangladesh.
34   *New Age*, 1 September 2014. *Bt* brinjal farmers demand
     compensation.
35   *New Age*, 21 March 2015. *Bt* brinjal turns out to be 'upset
     case' for farmers.
36   *New York Times*, 24 April 2015. How I Got Converted to
     G.M.O. Food.
37   Cornell Alliance for Science, 12 July 2016. Bangladeshi *Bt* brinjal
     farmer speaks out in GMO controversy, allianceforscience.
     cornell.edu/blog/bangladeshi-bt-brinjal-farmer-speaks-out-
     gmo-controversy.
38   *GM Watch*, 28 July 2015. Propaganda over facts? BBC
     Panorama and *Bt* brinjal, gmwatch.org/en/news/
     latest-news/16320.

39   *Marklynas.org*, 8 May 2014. *Bt* brinjal in Bangladesh – the true
     story, www.marklynas.org/2014/05/bt-brinjal-in-
     bangladesh-the-true-story/.

40   Mark Lynas, 14 May 2014. Bangladesh *Bt* brinjal farmers
     speak out, www.youtube.com/watch?v=_LoKPldPopU.

41   *Daily Inquirer*, 29 July 2016. Boost for *Bt* 'talong' opinion.
     inquirer.net/96038/boost-for-bt-talong.

42   *International Monsanto Tribunal*. Advisory Opinion, www.
     monsanto-tribunal.org/upload/asset_cache/189791450.pdf.

43   *IFOAM – Organics International*, 13 September 2016. People's
     Assembly & Monsanto Tribunal, www.ifoam.bio/en/
     news/2016/09/13/registration-open-peoples-assembly-monsanto-
     tribunal-14-16-october-2016-hague.

44   *Guardian*, 13 October 2016. GM seed firm Monsanto
     dismisses 'moral trial' as a staged stunt.

45   *International Monsanto Tribunal*. Advisory Opinion.

46   *International Monsanto Tribunal*. Progam – Monsanto Tribunal,
     www.monsanto-tribunal.org/program.

47   *ABC News*, 14 June 2014. GM farmer wins landmark canola
     contamination case in WA Supreme Court.

48   *Supreme Court of Western Australia*. Marsh vs. Baxter. 2014.

49   *International Monsanto Tribunal*. Memo no. 15 Farida
     AKTHER, www.monsanto-tribunal.org/upload/asset_
     cache/373558186.pdf?rnd=HknM44.

50   Kruger, M., et al. 2014. Detection of Glyphosate in
     Malformed Piglets. *Journal of Environmental and Analytical
     Toxicology*, 4: 5.

51   *EFSA*, 12 November 2015. Glyphosate: EFSA updates
     toxicological profile, www.efsa.europa.eu/en/press/
     news/151112.

52   *Reuters*, 18 April 2016. How the World Health Organization's
     cancer agency confuses consumers, www.reuters.com/
     investigates/special-report/health-who-iarc/.

53   *The Times*, 18 October 2017. Weedkiller scientist was paid
     £120,000 by cancer lawyers.

54   *Reuters*, 19 October 2017. In glyphosate review, WHO cancer
     agency edited out 'non-carcinogenic' findings.

55 *IARC.* IARC Monographs on the Evaluation of
Carcinogenic Risks to Humans, List of Classifications.
Volumes 1–119, monographs.iarc.fr/ENG/Classification/
latest_classif.php.

56 *New York Times*, 14 May 2015. Defying U.S., Colombia Halts
Aerial Spraying of Crops Used to Make Cocaine.

57 *Agronews.* Seven glyphosate companies listed first China's top
20 pesticide enterprises, news.agropages.com/News/
NewsDetail---10968.htm.

58 *International Monsanto Tribunal.* Memo no. 23: Claire
ROBINSON, www.monsanto-tribunal.org/upload/asset_
cache/328188625.pdf?rnd=E7bYWr

59 *The Guardian*, 3 February 2011. WikiLeaks: US targets EU
over GM crops.

60 *BBC News*, 7 January 2005. Monsanto fined $1.5m for
bribery.

61 *Wikipedia.* List of largest companies by revenue.

62 *Fortune*, 6 June 2016. Can Monsanto save the planet?

63 *Fortune 500*: Archive 1965, archive.fortune.com/magazines/
fortune/fortune500_archive/snapshots/1965/902.html.

64 *Oxfam International*, 26 April 2010. Oxfam International's
position on transgenic crops, www.oxfam.org/en/
campaigns/oxfam-internationals-position-transgenic-crops.

Chapter 6: Africa: Let Them Eat Organic

1 *UNICEF*, 10 April 2015. Survey shows sharp drop in
childhood stunting in Tanzania, www.unicef.org/media/
media_81517.html.

2 *BBC News*, 12 April 2007. Deaths in Uganda forest protest

3 *African Civil Society Statement: Call for a ban on GMOs* –
Acbio, org.za/activist/petition/African%20Civil%20
Society%20Statement%20Call%20for%20a%20ban%20
on%20GMOs.

4 Kenya Citizen TV, 21 November 2012, www.youtube.com/
watch?v=2qV75NOjsuY.

5 Seralini, G.-E., et al, 2012. RETRACTED: Long term
toxicity of a Roundup herbicide and a Roundup-tolerant

genetically modified maize. *Food and Chemical Toxicology*, 50,
11: 4221–4231.

6 *Food Sovereignty Ghana*, 8 July 2015. FSG Goes To Court
Today Over *Bt* Cowpeas and GM Rice, foodsovereignty
ghana.org/fsg-goes-to-court-today-over-bt-cowpeas-and-
gm-rice/.

7 *Food Sovereignty Ghana*, 20 May 2014. Ban All GM Foods
In Ghana! foodsovereigntyghana.org/ban-all-gm-foods-in-
ghana/.

8 *The Sunday Mail*, 8 June 2014. Mudede slams GMO academic,
www.sundaymail.co.zw/mudede-slams-gmo-academic/.

9 *Lusaka Times*, 7 June 2014. Lunanshya council destroys Bokomo
Cornflakes containing traces of GMO, www.lusakatimes.
com/2014/06/07/lunanshya-council-destroys-bokomo-
cornflakes-containing-traces-gmo/.

10 *UNICEF*, 2007. Nutrition in Zambia, www.unicef.org/
zambia/5109_8461.html.

11 *New York Times*, 30 August 2002. Between Famine and
Politics, Zambians Starve.

12 Paarlberg, R. 2009. *Starved for Science: How Biotechnology
is Being Kept Out of Africa*. Harvard University Press, US,
p. 15.

13 *Daily Telegraph* blogs. Why being Green means never having to
say you're sorry, web.archive.org/web/20101108023823/
blogs.telegraph.co.uk/news/jamesdelingpole/100062459/
why-being-green-means-never-having-to-say-youre-sorry/.

14 TVE Earth Report, 2005. Aliens in the Field, tve.org/film/
aliens-in-the-field/.

15 *The Guardian*, 17 October 2002. Zambians starve as food aid
lies rejected.

16 Greenpeace, 30 September 2002. Eat this or die: The poison
politics of food aid, www.greenpeace.org/international/en/
news/features/eat-this-or-die/.

17 Paarlberg, R. 2009. Ibid., p. 82.

18 Cornell Alliance for Science, 22 February 2017. Visiting
Tanzania's first-ever GMO crop trial, allianceforscience.
cornell.edu/blog/tanzania-first-ever-GM-maize-crop-
trial.

19 *Famine Early Warning System Net*, February 2017.

20 *Daily News*, 7 March 2017. Revoke GMO trials in Dodoma
    Mr President, www.dailynews.co.tz/index.php/analysis/48979-
    revoke-gmo-tech-trials-in-dodoma-mr-president.

21 *Little Atoms*, 19 April 2017. Tanzania is burning GM corn
    while people go hungry, littleatoms.com/science-world/
    tanzania-burning-GM-corn-while-people-go-hungry.

Chapter 7: The Rise and Rise of the Anti–GMO Movement

1   *Euractiv*, 2015. Jeremy Rifkin: 'Number two cause of global
    warming emissions? Animal husbandry', www.euractiv.com/
    section/agriculture-food/interview/jeremy-rifkin-
    number-two-cause-of-global-warming-emissions-animal-
    husbandry/.

2   Wade, N. 1973. Microbiology: Hazardous Profession Faces
    New Uncertainties. *Science*, 182, 4112: 566–567.

3   Watson, J. and Tooze, J. 1981. *The DNA Story: A documentary
    history of gene cloning.* W. H. Freeman and Company.
    Prologue.

4   Wade, N. 1973. Ibid.

5   National Academy of Sciences, 1977. *Research with
    Recombinant DNA: An Academy Forum*, March 7–9, 1977.

6   Watson, J. and Tooze, J. 1981. Ibid., p. 15.

7   Wade, N. 1973. Ibid.

8   National Academy of Sciences, 1977. Ibid.

9   Watson, J. and Tooze, J. 1981. Ibid., p. 14.

10 Watson, J. and Tooze, J. 1981. Ibid., p. 28.

11 Watson, J. and Tooze, J. 1981. Ibid., p. 43.

12 Watson, J. and Tooze, J. 1981. Ibid., p. 95.

13 *New York Times Magazine*, 22 August 1976. New strains of
    life – or death.

14 Watson, J. and Tooze, J. 1981. Ibid., p. 159.

15 Watson, J. and Tooze, J. 1981. Ibid., p. 160.

16 Watson, J. and Tooze, J. 1981. Ibid., p. 262.

17 Watson, J. and Tooze, J. 1981. Ibid., p. 169.

18 Watson, J. and Tooze, J. 1981. Ibid., p. 132.

19 Watson, J. and Tooze, J. 1981. Ibid., p. 235.

20 From Stewart Brand's *CoEvolution Quarterly*, Spring 1978, 17: 24.

21 *Pennsylvania Gazette*, October 1992. Jeremy Rifkin's Big Beefs.

22 *Pennsylvania Gazette*, October 1992. Ibid.

23 *Pennsylvania Gazette*, October 1992. Ibid.

24 Uhl, M., 2007. *Vietnam Awakening: My Journey from Combat to the Citizens' Commission of Inquiry on U.S. War Crimes in Vietnam*. McFarland & Co.

25 Application from the People's Bientennial Commission for a public gathering, 4 July 1976. Gerald R. Ford Presidential Library, www.fordlibrarymuseum.gov/library/document/0067/1563322.pdf.

26 *The Blade*, Toledo, Ohio, 22 April 1976. Backers of Revolutionary Concepts Stir Rebellion By Some in Business.

27 *Pennsylvania Gazette*, October 1992. Ibid.

28 Howard, T. and Rifkin, J. 1977. *Who Should Play God?* Dell Publishing Co. p. 10.

29 Howard, T. and Rifkin, J. 1977. Ibid., p. 44.

30 Howard, T. and Rifkin, J. 1977. Ibid., p. 206-7.

31 Howard, T. and Rifkin, J. 1977. Ibid., p. 224.

32 *Pennsylvania Gazette*, October 1992. Ibid.

33 *Pennsylvania Gazette*, October 1992. Ibid.

34 *The Gettysburg Times*, 16 November 1979. Author Warns Against Science 'Playing God'.

35 *Euractiv*, 2015. Ibid.

36 *New York Times*, 16 November 1986. Biotech's Stalled Revolution.

37 *BBC News*, 14 June 2002. GM crops: A bitter harvest?

38 *New York Times*, 25 January 2001. Biotechnology Food: From the Lab to a Debacle.

39 *The Washington Post*, 12 January 1993. Biotech tomato headed to market despite threats.

40 *New York Times*, Retro Report. Test Tube Tomato, www.nytimes.com/video/us/100000002297044/test-tube-tomato.html.

41 *The Washington Post*, 12 January 1993. Ibid.

42 Bruening, G. and Lyons, J., 2000. The case of the FLAVR SAVR tomato. *California Agriculture*, 54, 4: 6–7.

43 *New York Times*, 5 September 2015. Food Industry Enlisted Academics in G.M.O. Lobbying War, Emails Show.

44 Organic Consumers Association, www.organicconsumers. org/news/vaccine-studies-debunked.

45 Organic Consumers Association, www.organicconsumers.org/ news/ebola-can-be-prevented-and-treated-naturally-so-why-are-these-approaches-completely-ignored.

46 Organic Consumers Association, www.organicconsumers. org/categories/swine-bird-flu.

47 Charles, D. 2001. *Lords of the Harvest: Biotech, Big Money, and the Future of Food*. Basic Books.

48 Charles, D. 2001. Ibid., p. 100.

49 Charles, D. 2001. Ibid., p. 100.

50 Charles, D. 2001. Ibid., p. 208.

51 Charles, D. 2001. Ibid., pp. 208–9.

52 *Irish Times*, 13 March 1996. Attack on the mutant tomatoes a failure.

53 *Associated Press*, 14 February 2001. Europe OKs New Biotech Food Rules.

54 *CNN.com*, 8 February 2001. Bove on trial for wrecking genetic rice and CNN.com, 15 March 2001. Bove convicted for food assault.

55 *The Ecologist*, 29 January–1 February 1999. India cheers while Monsanto burns.

56 *St Louis Post-Dispatch*, 2 April 2001. Arsonists burn Monsanto depot in Italy.

57 Schurman, R. and Munro, W., 2010. *Fighting for the Future of Food: Activists Versus Agribusiness in the Struggle over Biotechnology*. University of Minnesota Press. Table 2, p. 108.

58 Schurman, R. and Munro, W. 2010. Ibid., p. 138.

59 National Center for Family Philanthropy, 2001. Practices in Family Philanthropy – Collaborative Grantmaking:

Lessons Learned from the Rockefeller Family's Experiences. National Center for Family Philanthropy, Washington D.C.

60 *Foundation for Deep Ecology.* Some Thought on the Deep Ecology Movement, www.deepecology.org/deepecology. htm.

61 *Foundation for Deep Ecology.* Work in Progress, www. deepecology.org/books/Work_In_Progress.pdf.

62 This information is gleaned from multiple tax returns. For a useful summary see archive.li/elmRO.

63 Greenpeace International, 2015. *Annual Report 2015,* www. greenpeace.org/international/Global/international/ publications/greenpeace/2016/2015-Annual-Report-Web.pdf.

64 *Academics Review,* 2014. *Organic Marketing Report,* academicsreview.org/wp-content/uploads/2014/04/AR_ Organic-Marketing-Report_Print.pdf.

65 Jay Byrne, Food & Agricultural Advocacy 2011–2012 Ag-biotech & GMO labeling case studies. Presentation, National Association of State Departments of Agriculture (NASDA), Des Moines, 2012, www.nasda.org/File. aspx?id=4275.

66 Friends of the Earth, 2015. *Spinning Food: How food industry front groups and covert communications are shaping the story of food,* www.foe.org/news/archives/2015-06-new-report-exposes-how-front-groups-shape-story-of-food.

Chapter 8: What Anti-GMO Activists Got Right

1 *The Observer,* 9 March 2013. Mark Lynas: truth, treachery and GM food.

2 *The Guardian,* 5 November 2010. Deep Peace in Techno Utopia, www.monbiot.com/2010/11/05/deep-peace-in-techno-utopia/.

3 *The Guardian,* 5 November 2010. George Monbiot's blog: When will Stewart Brand admit he was wrong? See also George's website, www.monbiot.com/2010/11/10/ correspondence-with-stewart-brand-second-tranche/.

4   *The Dark Mountain Manifesto*, dark-mountain.net/about/
    manifesto/.

5   Kingsnorth, P., 2011. *The Quants and the Poets*, paulkingsnorth.
    net/2011/04/21/the-quants-and-the-poets/.

6   Dawkins, R., 1998. *Unweaving the Rainbow: Science, Delusion
    and the Appetite for Wonder.* Penguin Books, London, p. 17.

7   Dawkins, R., 1998. Ibid.

8   Kuntz, M., 2012. The postmodern assault on science. *EMBO
    Reports*, 13, 885–889.

9   Oxfam America, 2015. *Land and Human Rights in Paraguay*,
    www.oxfamamerica.org/static/media/files/Paraguay_
    background.pdf.

10 Oxfam, 23 April 2014. *Smallholders at Risk: Monoculture
    expansion, land, food and livelihoods in Latin America*, www.
    oxfam.org/sites/www.oxfam.org/files/bp180-smallholders-
    at-risk-land-food-latin-america-230414-en_0.pdf.

11 *ETC Group*, 13 December 2016. Deere & Co. is becoming
    'Monsanto in a box', www.etcgroup.org/content/
    deere-co-becoming-monsanto-box.

12 Winner, L., 1986. *The Whale and the Reactor: A Search for
    Limits in an Age of High Technology.* University of Chicago
    Press, p. 9.

13 Mander, J. 1991. *In the Absence of the Sacred: The Failure of
    Technology and the Survival of the Indian Nations.* Sierra Club
    Books, p. 35.

14 Mander, J. 1991. Ibid., p. 27.

15 Thomas, J. 2008. Synthetic Biology Debate at the Long
    Now Foundation. longnow.org/seminars/02008/nov/17/
    synthetic-biology-debate/

16 Berry, W. Why I am not going to buy a computer, btconnect.
    com/tipiglen/berrynot.html.

17 *Wired*, 6 January 1995. Interview with the Luddite, www.
    wired.com/1995/06/saleskelly/.

18 *Wired*, 6 January 1995. Ibid.

19 Thomas, J. 21st Century Tech Governance? What would
    Ned Ludd do? 2020science.org/2009/12/18/thomas/.

20 Thomas, J. Ibid.

Chapter 9: How Environmentalists Think

1 *Kickstarter.com*. Glowing Plants: Natural Lighting with no
  Electricity, www.kickstarter.com/projects/antonyevans/
  glowing-plants-natural-lighting-with-no-electricit/
  description.
2 *ETC Group*, 7 May 2013. Kickstopper letter to Kickstarter,
  www.etcgroup.org/content/kickstopper-letter-kickstarter.
3 The American Chestnut Research and Restoration Project,
  www.esf.edu/chestnut/.
4 Fedoroff, N. and Brown, N.-M. 2004. *Mendel in the Kitchen:
  A Scientist's View of Genetically Modified Food*. National
  Academies Press, location 734.
5 Haber, J. 1999. DNA recombination: the replication
  connection. *Trends in Biochemical Sciences*, 24, 7: 271–275.
6 Directive 2015/412 of the European Parliament and of
  the Council of 11 March 2015 amending Directive 2001/18/
  EC as regards the possibility for the Member States to
  restrict or prohibit the cultivation of genetically modified
  organisms (GMOs) in their territory, eur-lex.europa.eu/
  legal-content/EN/TXT/HTML/?uri=CELEX:32015L0412
  &from=EN.
7 Center for Food Safety. GE Fish & the Environment, www.
  centerforfoodsafety.org/issues/309/ge-fish/ge-fish-and-the-
  environment.
8 Aquabounty.com. Sustainable, aquabounty.com/
  sustainable/
9 Haidt, J. 2012. *The Righteous Mind: Why Good People are
  Divided by Politics and Religion*. Penguin Books, London,
  p. 28.
10 Haidt, J. 2012. Ibid., p. 29.
11 Haidt, J. 2012. Ibid., p. 59.
12 Allow Golden Rice Now! The Crime against Humanity,
  allowgoldenricenow.org/wordpress/the-crime-against-
  humanity/.
13 Laureates Letter Supporting Precision Agriculture (GMOs),
  supportprecisionagriculture.org/nobel-laureate-gmo-letter_
  rjr.html.

14   Dawkins, R. 1998. Ibid., p. 31.

15   Cornell Alliance for Science, 23 May 2016. GMO safety
     debate is over, allianceforscience.cornell.edu/blog/mark-
     lynas/gmo-safety-debate-over.

16   Schulz, K., 2010. *Being Wrong: Adventures in the Margin of
     Error.* Granta Publications, p. 175.

17   Schulz, K. 2010. Ibid., p. 157.

18   Haidt, J. 2012. Ibid., p. 100.

19   Haidt, J. 2012. Ibid., p. 104.

20   Schulz, K. 2010. Ibid., p. 149.

21   Quoted in Schulz, K. 2010. Ibid., p. 152.

22   Schulz, K. 2010. Ibid., p. 156.

23   *BBC News*, 15 July 2010. Maldives atheist who felt persecuted
     'hangs himself'.

24   *Minivan News*, 9 June 2014.Vigilante mobs abduct young men
     in push to identify online secular activists, minivannewsarchive.
     com/politics/vigilante-mobs-abduct-young-men-in-push-
     to-identify-online-secular-activists-86720.

25   You'll find one of them here: www.eco-action.org/dod/
     no9/may_day.htm. Earth First! Journal *Do or Die*,
     issue 9. Note that I get a good slagging elsewhere in the
     journal!

Chapter 10: Twenty Years of Failure

1   ISAAA Brief 52-2016 – Executive Summary, www.isaaa.org/
    resources/publications/briefs/52/executivesummary/default.
    asp. This is 185 million hectares out of roughly 1.5 billion
    global total.

2   *Hawaii News Now*, 29 September 2009, www.huffingtonpost.
    com/2013/09/29/ecoterrorism-papayas-hawaii_n_4013292.
    html.

3   Greenpeace International, 27 July 2004. GE papaya scandal
    in Thailand, www.greenpeace.org/international/en/news/
    features/ge-papaya-scandal-in-thailand/.

4   Lynas, M. and Evanega, S.-D., 2015. The Dialectic of
    Pro-Poor Papaya. In Ronald J. Herring (ed.), *The Oxford*

Handbook of Food, Politics, and Society. Oxford University Press, Oxford.

5   Davidson, S., 2008. Forbidden Fruit: Transgenic Papaya in Thailand. *Plant Physiology* 147: 487–493.

6   Greenpeace, 31 August 2012. 24 children used as guinea pigs in genetically engineered 'Golden Rice' trial, www. greenpeace.org/eastasia/news/blog/24-children-used-as-guinea-pigs-in-geneticall/blog/41956/.

7   Klumper, W. and Qaim, M. 2014. A Meta-Analysis of the Impacts of Genetically Modified Crops. *PLOS One*, 9, 11: e111629.

8   Brookes, G. and Barfoot, P., 2017. Environmental impacts of genetically modified (GM) crop use 1996-2015: Impacts on pesticide use and carbon emissions. *GM Crops & Food*, 8, 2: 117-147.

9   *Union of Concerned Scientists*. Environmental impacts of coal power: air pollution, www.ucsusa.org/clean-energy/coal-and-other-fossil-fuels/coal-air-pollution.

10 End Coal. Coal Plants by Country (units), endcoal.org/wp-content/uploads/2017/07/PDFs-for-GCPT-July-2017-Countries-Units.pdf.

11 Greenpeace International. Why we must quit coal, www. greenpeace.org/international/en/campaigns/climate-change/coal/.

12 National Academy of Sciences, 2016. *Genetically Engineered Crops: Experiences and Prospects*. Washington, D.C.: The National Academies Press, p. 96.

13 Environmental Defense Fund. Monarch Butterfly Habitat Exchange, www.edf.org/ecosystems/monarch-butterfly-habitat-exchange.

14 Lu., Y., et al. 2012. Widespread adoption of *Bt* cotton and insecticide decrease promotes biocontrol services. *Nature*, 487, 7407: 362–365.

15 National Academy of Sciences, 2016. Ibid. p. 98

16 Hilbeck, A. et al, 2015. No scientific consensus on GMO safety. *Environmental Sciences Europe*, 27: 4.

17 Global Warming Petition Project. www.petitionproject.org/.

18 Discovery Institute, 24 September 2001. 100 Scientists,
    National Poll Challenge Darwinism, www.reviewevolution.
    com/press/pressRelease_100Scientists.php

19 National Center for Science and Education, ncse.com/
    project-steve-faq.

20 Genetic Literacy Project. Jeffrey Smith: Former flying yogic
    instructor now 'most trusted source' for anti-GMO
    advocacy. geneticliteracyproject.org/glp-facts/
    jeffrey-m-smith/.

21 Pew Research Center, 1 December 2016. Public opinion
    about genetically modified foods and trust in scientists
    connected with these foods. www.pewinternet.
    org/2016/12/01/public-opinion-about-genetically-
    modified-foods-and-trust-in-scientists-connected-with-
    these-foods/.

22 Pew Research Center, 29 January 2015. Public and Scientists'
    Views on Science and Society. www.pewinternet.
    org/2015/01/29/public-and-scientists-views-on-science
    -and-society/.

23 European Commission, 2010. *A Decade of EU-funded GMO
    Research (2001–2010)*, ec.europa.eu/research/biosociety/
    pdf/a_decade_of_eu-funded_gmo_research.pdf.

24 Environmental Defense Fund. Our position on
    biotechnology, www.edf.org/our-position-biotechnology

25 Purdue University, 29 February 2016. Study: Eliminating
    GMOs would take toll on environment, economies,
    www.purdue.edu/newsroom/releases/2016/Q1/study-
    eliminating-gmos-would-take-toll-on-environment,-
    economies.html.

26 Taheripour, F., et al, 2016. Evaluation of economic, land use,
    and land-use emission impacts of substituting non-GMO
    crops for GMO in the United States. *AgBioForum*, *19*, 2:
    156–172.

27 FAO, 2015. *Global Forest Resources Assessment 2015*. www.fao.
    org/3/a-i4793e.pdf p. 3

28 Renewable Fuels Association. Industry Statistics, www.
    ethanolrfa.org/resources/industry/statistics/.

29 USDA, 2017. *U.S. Bioenergy Statistics*, www.ers.usda.gov/
    data-products/us-bioenergy-statistics/.

30 Greenpeace Finland, 23 February 2011. Research on palm oil
    and biofuels, www.greenpeace.org/finland/en/What-we-do/
    Neste-Oil--driving-rainforest-destruction/Research-on-palm-oil-
    and-biofuels/

31 Wilson, E., 2016. *Half-Earth: Our Planet's Fight for Life.* Liveright
    Publishing.

32 Monbiot, G., 2013. *Feral: Searching for Enchantment on the
    Frontiers of Rewilding.* Penguin. p. 153.

Acknowledgements

A lot of people helped with this book, often in quite unexpected ways. Some of them wish to remain anonymous, but are no less deserving of thanks for that. One in particular provided feedback that was nothing short of transformative. (You know who you are!)

I am particularly grateful to those who offered their time as well as their expertise in helping me with research as well as numerous rewrites. Thanks in particular go to Marc Van Montagu and Nora Podgaetzki for their kindness and generosity in both putting me up in Brussels and sharing some of their fascinating life stories. I also thank Mary-Dell Chilton and Robb Fraley for likewise kindly sharing their recollections about the process of the invention of genetic engineering.

I am also very grateful to Jim Thomas for his trust, integrity and honesty as we recalled our early exploits and explored current common ground we still might share. George Monbiot was as always exceptionally generous with his time and experience, and persevered with helping me understand issues of political economy that 'sciencey' types like me often prefer to ignore. A special mention must go to Paul Kingsnorth, together with whom I have travelled long and far in this life. Our paths diverged for a while, but seem to have come together again, and for that I am truly grateful. Maybe we can be friends without agreeing on everything after all. Paul's writing skills are second to none, and his comments and suggestions on drafts of this book were invaluable.

Alison Van Eenennaam, who is both a stellar scientist and a gifted communicator, kindly provided comments too, as did the equally stellar Pam Ronald. Both are to be found at UC Davis. I am also grateful to Nina Fedoroff, who helped inspire me to look more deeply into the 'GMO' issue when I took my first fumbling steps of rediscovery back in 2013.

Having the institutional basis of the Alliance for Science at Cornell University was invaluable to me between 2014 and 2017, and many of the stories I tell in this book came from research and travel I undertook with Cornell. Thanks, in particular, to Sarah Evanega and Joan Conrow there, both of whom read and kindly commented on drafts. The Alliance for Science in turn would not have become the success it is without the enduring support of the Bill & Melinda Gates Foundation, whose program officers and others have helped with much more than just funding. Although this book has been an independent project of mine, for which I retain full responsibility, I have benefited hugely from these relationships.

I would also like to thank Stewart Brand and Ryan Phelan at Revive & Restore, for being something of a foil in this book but also for their wisdom and vision. Stewart's book *Whole Earth Discipline* was a special inspiration for me and deserves a special mention. Matt Ridley also kindly sent comments on a first draft, as did Paul Roberts, my co-conspirator from the old Maldives days. So did the filmmaker Robert Stone, director of Pandora's Promise – where, incidentally, you can see me visiting Fukushima soon after the accident. Over in sunny Hawaii I would like to thank Rory Flynn, whose deep dives into the tortuous funding connections of anti-GMO groups provided invaluable information, much of which I have not had the space to include here.

I would also like to thank Tim Harford, who – one spring morning over coffee – breezily suggested that I should stop dawdling and just get on and write the thing. I'm also grateful to Charlotte Croft for our subsequent serendipitous chat in the school playground. My agent Antony Harwood immediately got the concept and provided invaluable feedback, expertise and support throughout, as he has done consistently since my very first book way back in 2004. I'm especially grateful to Jim Martin and Anna MacDiarmid at Bloomsbury for taking this book on and investing so much time into it, as well as to copy-editor Catherine Best for helping improve dramatically on my first efforts.

I would also like to mention my friends and neighbours in Wolvercote, especially Paul and Joan Rimmer (93 years young!), Nigel, Louise, Zeb, Mel, Dave, Teresa, Lucy, Alex and Didier Delgorge, all the Kirstie Mortons (okay, ex-Wolvercote), and everyone at my two 'locals', The Plough and Jacobs Inn. Suzanne at the latter deserves special credit for being a smiling face and helping with the dogs. It takes a village to raise a child, and also to write a book it seems. Especially a village with good pubs.

Deepest and most profound thanks of course go to my family. My beautiful children Tom and Rosa had to put up with quite a few times when Dad was even more grouchy than normal. My wife Maria is both my number one supporter and number one critic, managing to combine these roles with love, intelligence and sensitivity. I literally could not have done this without her, whatever she might say. My wonderful parents, Val and Bry Lynas, also contributed in so many ways – in particular my father as we took this journey together, me as a writer, him as an organic farmer.

I dedicate this book to the memory of David MacKay, who was both a friend and a mentor to me, and who helped impress on me – as someone who was always a bit hopeless at maths – the unmatchable importance of numbers. We shared a love of both empiricism and the Proclaimers. Here's to the next 500 miles!

Index